Contents

Foreword

The planning of roads and footpaths has a major impact on the character, quality and safety of residential areas. It affects the movement of traffic, including emergency and other services, and arrangements for the parking of residents' and visitors' cars. It affects the ease with which people, including those who are disabled, can get around. It affects the layout of houses and their gardens and provision for children's play and other outdoor activities.

This new edition of Design Bulletin 32 is the result of detailed consultation with a wide variety of interests. We hope that the information and advice it contains will help everyone involved with the design of new housing developments, and the renovation of older residential areas and estates, to create an environment which is attractive, convenient, safe and economical to construct and maintain.

Standards for road and footpath layout which result from applying the advice in this bulletin will allow a sensible balance to be struck between planning, housing and highway considerations in the design of most new residential developments and improvement schemes. We expect local planning and highway authorities to take the advice contained in this edition into account when preparing local guidance and when specifying their requirements for adoption.

Sir George Young Bt MP
Minister for Housing and Planning, Department of the Environment

Kenneth Carlisle MP
Minister for Roads and Traffic, Department of Transport

Introduction

Purposes and scope of this edition

Design Bulletin 32[1] was published in 1977 and has been widely used ever since as a reference on the layout of roads and footpaths in new residential development. The main purposes of this second edition are to update and amend the information and advice in the original bulletin in the light of experience of its use and changes in housing over the last 14 years, take into account new initiatives on road safety[2] and make the bulletin more relevant than before to the planning of improvement schemes in existing residential areas and older public sector housing estates. Advice in this edition supersedes that contained in the first edition.

Section 1 describes the main considerations that need to be taken into account when producing a design brief for the layout of roads and footpaths. It contains new guidance on developing an appropriate overall design concept - taking into account the policies and proposals in development plans, the characteristics of the site and its setting; arrangements for landscape maintenance; the various functions performed by roads and footpaths; the access requirements of pedestrians, cyclists and drivers; considerations of road safety, traffic nuisance and security from crime and vandalism, and the characteristics of the road system around the development site.

Section 2 discusses the overall layout of roads and footpaths. It includes new guidance on means to minimise danger and nuisance from non-access traffic, reduce vehicle flows, restrain vehicle speeds, provide safe, convenient and secure routes for pedestrians and cyclists and make effective provision for parking.

Section 3 considers the detailed design of each element in the layout - the carriageways, junctions, turning spaces, footways, verges, footpaths and parking areas - and requirements for intervisibility. It includes new guidance on facilities for pedestrians and cyclists, planting in verges and dimensions for parking areas.

As in the first edition, no attempt is made to prescribe standards for the adoption of highways or for the control under planning powers of the layout of new residential roads and footpaths. This can only be done sensibly locally. However, Section 4 offers some fresh general advice on this topic and on the preparation of local standards for parking provision. Also, Sections 2 and 3 offer new guidance on the standards that will normally be appropriate for the layout of roads and footpaths in new developments.

As before, the guidance refers to relevant empirical evidence and to evidence drawn from the experiences of those engaged in practice. Also, again, the bulletin deals mainly with principles rather than design solutions. Some of these principles, particularly those for restraining vehicle speeds, take into account experience from abroad. Some have only been applied in this country for a short time and empirical evidence and experience of their use is limited. The encouragement of innovation has been balanced with caution where risks to safety may be involved.

Section 5 describes the special considerations that need to be taken into account when using the information and advice presented in Sections 1-3 to develop design briefs and plan improvements to the layouts of existing residential roads and footpaths.

Since the first edition was published, the National Joint Utilities Group (covering the gas, water, electricity and telecommunications industries) has produced comprehensive guidelines on procedures and technical requirements for the installation and location of buried services.[3] Consequently, less detailed information on this subject is given in this edition.

Requirements for distributor roads, matters of construction specification and guidance on parking controls such as waiting restrictions remain outside the scope of this bulletin.

Definitions for the purposes of this bulletin

The following definitions have been assumed for the purposes of this bulletin.

The urban road network

'*Primary distributors*' form the primary network for the town as a whole and all longer-distance traffic movements to, from and within the town are canalised on to such roads.

'*District distributors*' distribute traffic between the residential, industrial and principal business districts of the town and form the link between the primary network and the roads within residential areas.

'*Local distributors*' distribute traffic within districts. In residential areas, they form the link between district distributors and residential roads.

'*Residential access roads*' link dwellings and their associated parking areas and common open spaces to distributors. Such roads are referred to in this bulletin as residential roads.[4]

Residential roads and driveways

'*Access roads*' are residential roads with footways that may serve up to around 300 dwellings and provide direct access to dwellings (see Paragraph 2.13).* Where minor or major access roads are referred to it is assumed that they may serve up to around 100 and 100-300 dwellings respectively.

'*Shared surface roads*' are residential roads without footways that may serve up to around 50 houses (see Paragraph 2.70).

'*Shared driveways*' are unadopted paved areas that may serve the driveways of up to 5 houses (see Paragraph 2.81).

'*Driveways*' are unadopted paved areas that provide access to garages and other parking spaces within the curtilage of an individual house.

Other definitions

'*Carriageways*' are those parts of access roads which are intended primarily for use by vehicles.

'*Shared surfaces*' are those parts of shared surface roads which are intended for use by both pedestrians and vehicles.

'*Footways*' are those parts of access roads which are intended for use by pedestrians and which generally are parallel with the carriageways and separated by a kerb or verge and a kerb.

'*Footpaths*' are those pedestrian routes[5] which are located away from carriageways and not associated with routes for motor vehicles.

'*Cycle tracks*' are routes which are intended for use by pedal cyclists, with or without rights of way for pedestrians.

'*Segregated cycle tracks*' are cycle tracks adjacent to footways or footpaths, and separated from them by a feature such as a kerb, verge or white line.

---

* A factor of one vehicle journey per dwelling in the peak hour has been assumed for these definitions and elsewhere in this bulletin where suggested standards are related to the numbers of dwellings served by a road. When reference is made to the number of dwellings served by a road it should be borne in mind that the road may carry vehicular traffic not only from the dwellings that are located along its length but also from the dwellings served by any roads which branch off it. The largest vehicle flow in a cul-de-sac road will occur close to its entrance. For a loop or through road, it may normally be reasonable to assume that vehicle flows will be divided equally between entrances at each end.

The Design Brief

Main aims and objectives

1.01 Differences between residential and other types of roads are not only of scale; considerations are involved which do not apply elsewhere in the urban road system. Residential roads and footpaths are an integral part of housing layout where visually attractive and well-maintained surroundings which are secure and free from traffic nuisance are of prime importance and where in the patterns of movement around buildings the needs of pedestrians and cyclists for safety and convenience are given priority in design over the use of motor vehicles. To meet these aims in a new residential development it is necessary at the outset to develop an appropriate design brief for the road and footpath layout.

1.02 The requirements which have to be taken into account in the brief must first be established with local planning and highway authorities, those who provide statutory and other services, the fire and ambulance services and police advisers on crime prevention.

1.03 The brief should ensure that the road and footpath layout can be designed to suit the overall design concept for the development - taking into account:

(a) requirements and preferences in local development plans and policies and the housing development brief with regard to matters such as:

   (i)   the visual character of the development;

   (ii) building density;

   (iii) dwelling sizes and types;

   (iv) access for pedestrian and cyclists;

   (v)   access for vehicles, including emergency services and public transport;

   (vi) parking accommodation;

   (vii) private and common open space;

   (viii) daylight, sunlight, privacy and views;

   (ix) provision for trees and other forms of planting;

   (x)   landscape maintenance.

(b) the physical characteristics of the site and its surroundings;

(c) the location of safety hazards and traffic nuisances along nearby roads;

(d) the incidence of crime and vandalism in the local area;

(e) the volumes and types of pedestrian, cyclist and vehicular traffic likely to be generated by the scheme and any links with other planned developments - taking into account matters such as:

   (i)   the location of places in the surrounding area that will attract pedestrians and cyclists - such as bus stops, shops, schools, parks and work places;

(ii) the main directions in which vehicular traffic is likely to go when moving between the homes and destinations outside the site;

(iii) the possibility of providing special links for buses;

(iv) the intended functions of the surrounding roads and the likelihood of non-access vehicular traffic wanting to use the site to take short cuts;

(v) existing or proposed speed limits for roads within and around the site.

Development plans

1.04 Development plans set out the policies and proposals for the development and other use of land in an area, including those for the improvement of the environment and the management of traffic.[6] Such plans will provide the framework within which the design of the residential road and footpath layout should be set. Structure plans (in the shire counties) and part I of Unitary Development Plans (in the former metropolitan counties and Greater London respectively) provide the wider planning context.

1.05 Local plans, and part II of Unitary Development Plans, set out the development control policies and planning concepts applicable to the area which should be followed by designers. For example, the plan may show the circulation and distributor road system the local authorities wish to see operate around the site and the general form of building development they expect to relate to this system. The local plan is the medium through which local authorities can establish the hierarchy of roads for an area, other than the primary network which is a structure plan (and part I of an UDP) matter. Residential roads form part of the hierarchy and it is important that they are consistent with and complementary to the other levels.

Supplementary planning guidance

1.06 In addition to the requirements of statutory local plans, local authorities may produce supplementary planning guidance on matters of layout, consisting of general advisory and educational material to stimulate better design. It is desirable that such guidance be produced in consultation with house-builders and other interested bodies. Advisory material of this kind, such as design briefs, should be issued separately from the development plan, with its status made clear. When such guidance has been produced it will need to be taken into account when developing a design concept for the overall form of a development.

Overall design concept

Site characteristics and general housing requirements

1.07 For most new developments, a site survey is an essential early step in the development of an appropriate design concept. The constraints and opportunities identified have to be considered in relation to a wide range of housing requirements that may affect the layout of roads and footpaths. For instance:

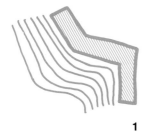

1

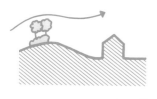

2

(a) the economics of developing a steeply sloping site is normally a major influence on the layout - with the roads and buildings having to follow the contours wherever possible (Figure 1). Even gentle slopes may affect the vertical alignment of roads in relation to adjacent land and buildings - and hence the landscape character of the development;

(b) exposure to wind, rain and snow may also be a major influence depending upon the location and topography of the site and its surroundings. The layout of buildings, roads and footpaths and landscape features such as hedges and trees can be used to ensure that spaces around buildings obtain maximum benefit from fine weather and some protection from adverse weather (Figure 2);

(c) to aid the conservation of energy in dwellings, it may be appropriate to align the roads to allow wide, south facing frontages; plant coniferous trees to the north and deciduous trees to the south of the houses and choose building forms that limit overshadowing (Figure 3).[7]

(d) to help enhance security, layouts should normally be designed to allow natural surveillance of roads from dwelling windows (see Paragraphs 1.54 and 2.18) - with any necessary protection from traffic noise along busy distributor roads being provided by means such as double-glazing. However, in exceptional circumstances it may be appropriate to design the layout to accommodate noise barriers (Figure 4) or the dwellings to face away from the noise (Figure 5).[8]

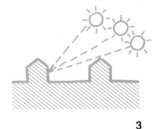

3

4

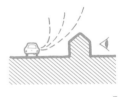

5

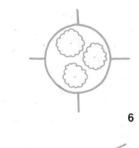

6

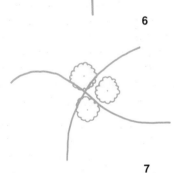

7

1.08 The need to preserve existing planting may also be a major influence on the layout (Figures 6 and 7). A detailed survey can establish the condition and other characteristics of existing trees and shrubs and identify those that should, if possible, be retained because they can adapt to the new environment created by the development and are likely to continue to grow. Such planting can add variety to the landscape and reduce the visual impact of carriageways and parking areas when the development is first occupied. Also, together with new planting, it can provide views that change from one season to another, give privacy and shelter from wind and filter dust from polluted air.

1.09 The proposed building density and the preferred types of accommodation are major considerations to be taken into account when appraising the likely feasibility of alternative design concepts - for they will influence the form of the development and the amount of space that can be provided around buildings. The amount of space that needs to be provided around buildings will be influenced in turn by the preferred design concept - in particular by whether the aim is to create a scene that is dominated by buildings and paved surfaces or by trees and other plants (Figures 8 and 9).

8

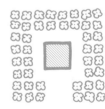

9

1.10 Guidance in this bulletin aims to allow sufficient space to be created in the layout to accommodate trees - by ensuring that roads and footpaths do not occupy space unnecessarily. But the space to be occupied by dwellings may also be relevant. For example, existing trees may be readily retained and new trees planted when detached houses are to be used at relatively low building densities, whereas at relatively high building densities it may only be possible to achieve this end by using terraced houses or flats (Figures 10-12). The overall design concept, the building density and the types of accommodation proposed should all be compatible.

10

11

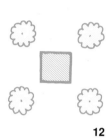

12

Landscape maintenance

1.11 Specialist advice should normally be sought when undertaking the site survey and analysis and when producing the landscape design, protecting existing features, repairing any damage that occurs during building, preparing the ground for planting, installing new trees and shrubs and producing a long-term maintenance plan.

1.12 The design concept for the development must take into account local policies for the adoption and maintenance of planted areas and the likely adequacy of maintenance arrangements. Normally, adequate maintenance in the long-term has to be assured by designing the layout so that areas provided for the general public benefit can be adopted by the local authority - with the developer making provision for the maintenance of areas provided for the benefit of the development itself.[9]

1.13 When considering their policies for adoption, local authorities need to recognise that common open spaces, verges and incidental areas of paving in the road and footpath layout may be the only places available in many developments where trees - especially the larger growing species of trees - could be planted (or preserved) without overshadowing dwellings. Developers need to recognise that planted areas for public adoption should be designed for minimum maintenance once handed over to the local authority.

1.14 Special care in design is needed to minimise risks of damage to planted areas and paving from pedestrians and vehicles, and to ensure that wear which is inevitable is an unobtrusive and acceptable part of the overall scene. Risks of damage by over-running vehicles will mainly be reduced by making adequate provision for off-street and on-street parking and by ensuring that carriageways are wide enough to allow vehicles to pass each other.

1.15 But the location and design of areas for planting, the selection of appropriate plant species and the choice of paving materials and detailing can also make an important contribution. For instance, low fencing or hedges may be needed to separate footways or footpaths from gardens or common open spaces - to help limit damage which can occur when children are at play or wear and tear caused by pedestrians taking short cuts. Fencing and plant species would need to be sufficiently robust to withstand such activities.

Road and footpath functions

1.16 Though this bulletin mainly gives guidance on the considerations that need to be taken into account when designing for the movement of pedestrians, cyclists and drivers, it recognises that residential roads and footpaths also perform other functions. They affect daylight, sunlight and privacy for the dwellings and gardens they serve, provide routes for statutory and other services and offer opportunities outside the home for children to play and neighbours to meet. They are also major elements in the overall scene and must be designed with skill and imagination to help create visually attractive surroundings.[10]

1.17 The principles set out in Section 2 of this bulletin have been developed to enable a wide range of design solutions to be produced. A great variety of road and footpath arrangements are needed to suit different design concepts for different kinds of sites and urban, suburban and rural settings - for instance:

(a) rectilinear configurations to help create formal surroundings in which the roads are prominent elements in the overall scene (Figure 13);

(b) curvilinear and other informal configurations to help create picturesque surroundings in which the roads are unobtrusive (Figure 14);

(c) various kinds of streets, avenues, crescents, squares and courts with geometrically regular layout forms (Figures 15-18);

(d) roads such as closes and lanes with geometrically irregular forms (Figures 19 and 20);

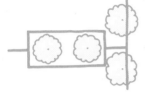

13

14

15

16

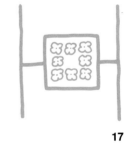

17

18

19

20

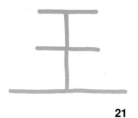

21

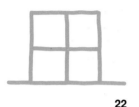

22

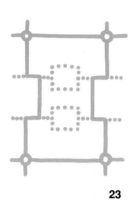

23

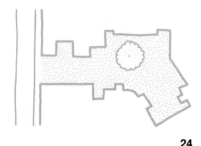

24

(e) hierarchical layouts with tree-like or network configurations (Figures 21 and 22) or other geometrically more complex arrangements.

1.18 Design concepts for road and footpath layouts need to embrace all elements that can be seen along the routes taken when entering a development, when moving around within it and when looking out from each home. All elements need to be considered together and be designed so that they are compatible. All have an important part to play in creating attractive roads and footpaths - the dwellings, garages, driveways, gardens, paths, gates, fences, footways, verges, common open spaces, carriageways, edge restraints, paving materials, street furniture, shrubs and trees.

1.19 For large developments, the overall design concept should indicate to drivers the different functions of different types of roads and thereby help pedestrians, cyclists and drivers who are strangers to find their way around. For instance, the different functions of major and minor access roads may be indicated by differences in the means that are used to restrain vehicle speeds (Figure 23) (see Section 2), layout configurations, building heights, and forms of parking provision. Shared surface roads need to be clearly different from other roads for an additional reason - to help indicate that drivers, pedestrians and cyclists are intended to share the same surface. This can be readily achieved when the carriageway alignment is closely integrated with the configuration of dwellings (Figure 24). Other means are suggested in Section 2.

1.20 Differing functions may be further emphasised by the selection of trees and other plants.[11] For example, broad crowned trees and grassed verges may be appropriate in the widest spaces along major access roads (Figure 25) and fastigiate trees and ground cover shrubs along minor access and shared surface roads (Figure 26). Climbing shrubs on buildings and elsewhere within dwelling curtilages may also play a part - especially where space on the ground is restricted or other forms of planting might adversely affect safety (by obstructing visibility) or security (by providing shelter for criminals). Carefully designed differences in paving materials, edge restraints and lighting arrangements may also be used to help indicate the functions of different roads. Planting and changes in pavings may be used as complementary measures to help restrain driving speeds (see Section 2).

25

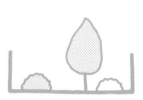

26

Access facilities

Modes of travel

1.21 Whilst the location of the development and the sizes and types of dwellings to be provided will influence the numbers of households that will own cars, the popularity of different modes of travel cannot be predicted with any certainty and in any case will change from time to time. There will also be the needs of visitors and the emergency and other services to take into account. Consequently most parts of most developments will need to provide access facilities for pedestrians, cyclists and drivers.

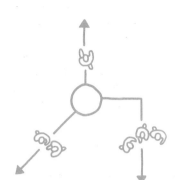

27

28

Pedestrians

1.22 Pedestrians include the very young and the very old, pram and wheelchair pushers, ambulant disabled people, wheelchair users and people with impairments of sight and hearing. Most residents are pedestrians for at least some of their journeys and require the opportunity to use safe, direct, secure and visually attractive routes to destinations such as local bus stops, shops, schools, parks and other community facilities. The need for such routes is of over-riding importance for children and for adult residents who do not own cars.

1.23 Pedestrian focal points in the local area, places where pedestrian safety is at risk and the location of any barriers for wheelchair users and other people with disabilities should be identified at the outset (Figure 27). Improvements in the area around the site may need to be considered (see Paragraph 1.49). Also, in the design of the development, special efforts may be needed to improve facilities for pedestrians in specific places - for example by the provision of shelter and passenger information facilities at a bus stop.

1.24 Children move frequently from one play space to another, and journeys to local community facilities are often a part of their play patterns, especially when they are unaccompanied by adults. It is common for residential roads to be crossed frequently by children, and for children to play on carriageways and in parking areas regardless of whether special play facilities are provided on the estate or in the area around.[12] Likely patterns of movement and play need to be taken into account when designing all parts of the road and footpath layout.

1.25 Also, though it may be technically impossible to design new developments to meet all the needs of elderly and disabled people - to ensure, for example, that there are no difficult gradients for wheelchair users or that bus stops can be reached by all those with severely limited walking ability - a variety of practical arrangements that can be of positive benefit are described in Sections 2 and 3. Additional information is available in other publications.[13]

Cyclists

1.26 Cyclists are one of the most vulnerable groups of road users, particularly young children who ride bicycles when out at play around the home and when going from their homes to schools and other local community facilities. Cyclists of all ages need safe, direct and secure routes. Focal points for residents who are cyclists should be identified at the outset, bearing in mind that some of their destinations may be further away than local community facilities (Figure 28).

1.27 Sections 2 and 3 suggest ways in which residential roads and footpath links may be designed to provide safe and convenient routes for both pedestrians and cyclists. However, when a development needs to be integrated into a wider system of provision for cyclists, it may be necessary to consider cycle tracks contiguous with but segregated from footpaths or footways. Such provision is primarily a matter for local authorities to decide in light of their overall transportation planning and considerations of cyclists needs and safety.[14]

Drivers

1.28 Though pedestrians and cyclists are the most vulnerable road users, and their needs have to be given priority in the layout of residential roads and footpaths, it is also essential to provide residents with facilities for vehicular access and to make effective provision for parking vehicles.

1.29 It cannot be assumed that all residents will have a car, nor that those who do will use a car for all journeys. Public transport usage could become more significant, particularly for local journeys. Most residents will need to use taxis from time to time and will want to have these vehicles come to the door. Also, the needs of elderly, frail and disabled people should be considered. These groups will prefer easy access to conventional bus services, but may also be able to benefit from dedicated services provided by minibuses, some of which will need to operate door-to-door. In some cases, requirements for a regular bus service and access for school and work buses will need to be considered.

1.30 Experience suggests that car owners also normally wish to have direct vehicular access to their homes. An unfortunate consequence of many past attempts to improve general amenity and safety by separating pedestrians from vehicles in the immediate vicinity of the home has been to decrease the convenience and security of residents' parking spaces. Residents have gone to considerable lengths to get their cars to the front door - to the extent of driving along footpaths and over greens - and cars parked out of view have been especially prone to crimes of theft and vandalism.

1.31 Fire, ambulance and other emergency services and public transport services have also found carriageways blocked by parked vehicles in schemes where both inadequate off-street parking provision is made and the carriageways are too narrow. It is in these situations that parking on footways has caused the greatest damage, and also hazards for visually handicapped, elderly and disabled people.

1.32 The aim should be to provide direct vehicular access to dwellings and a sufficient number of well-located parking spaces to minimise risks to pedestrian safety and the adverse cost and visual consequences of damage to footways and verges caused by indiscriminate on-street parking and over-running vehicles. Equally, excessive off-street or on-street parking provision wastes land and other resources and can adversely affect the visual character of developments. Sections 2-4 give guidance on designing and setting standards for parking provision.

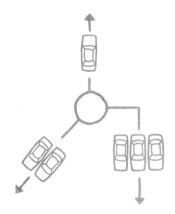

29

1.33 For large developments, likely destinations outside the site for residents who are car owners need to be identified at the outset (Figure 29). The aim should be to provide the shortest practicable routes to these destinations (see also Paragraph 1.48).

Public transport

1.34 Whenever possible, all schemes should be directly served or be within easy reach of a bus network, and new development can sometimes help to make this possible for existing or proposed new housing that would otherwise be poorly served.[15] To achieve this, close co-operation is required between public transport operators, the local authority* and the developer. This cooperation should continue from the initial planning stage to the completion of development.

1.35 Since deregulation of the bus industry under the Transport Act 1985, bus operators outside London are free to run services wherever they see a commercial opportunity and the scheme should take that possibility into account.[16]

---

* Appropriate authorities are the County Councils in the Shire counties; the Passenger Transport Authorities and Passenger Transport Executives in metropolitan areas outside London; and London Regional Transport in London.

1.36 The roads that are likely to be used by buses should be identified at the outset. Residents get into the habit of using buses if the service is operational when they first move in. Bus operators should be encouraged to start their services as soon as there are enough residents to outweigh the economic and practical disadvantages of serving a partly occupied development. If none of the operators is prepared to run commercial services initially because of uncertainty about demand, the local authority may wish to use its powers under the 1985 Act to subsidize the running of services under contract until demand picks up as the scheme develops. Large developments should be planned to allow the earliest phases to be provided with a bus service.

1.37 The use of a residential road as a bus route does not necessarily warrant its designation as a distributor nor is there necessarily a need to restrict direct access to dwellings. Many buses today are no larger than the service vehicles which normally use residential roads (see Section 3). Consequently, such buses may use roads designed in accordance with the recommendations set out in this bulletin. The special considerations which need to be taken into account in these circumstances are outlined in Section 2. It will normally be appropriate for larger vehicles to use distributor roads. Such vehicles may also use major access roads when this would be the only way to ensure an effective bus service. Design requirements for these roads should be determined in consultation with local public transport operators.

Danger and nuisance from traffic

Accidents

1.38 Outside town centres, roughly half the accidents in built up areas occur on main (arterial) routes.[17] A further quarter occur on local distributors which often give direct access to dwellings, and are crossed and followed by pedestrians travelling between the home and destinations such local community facilities, schools and places of work. The remaining quarter of accidents occur in residential areas, although residential roads free from non-access traffic are unlikely to account for more than 10 per cent of urban accidents. Junctions providing access from main distributor roads to residential areas are vulnerable points and need to be carefully planned for safety.

1.39 An outstanding feature of accidents in residential areas is the high involvement of child pedestrians. The proportion of child to adult casualties is highest on roads which are serving only as access to dwellings. Half of all road accidents to children under five years occur within 100m of their homes.[18] The government's strategy for reducing road accidents involving children is set out in 'Children and Roads: A Safer Way'.[19] One part of this strategy is to make the roads safer.

1.40 Surveys show that very few accidents occur in culs-de-sac and short loops that function only as residential roads. Culs-de-sac serving up to 80 dwellings were included in the surveys and there was no statistically significant increase in accident rate per dwelling associated with the size of cul-de-sac at least up to that size. Traffic flows are normally low on these roads, as are vehicle speeds.[20] Also, it has been shown that reducing speeds to 20mph and below has a significant effect on the number of pedestrians killed and seriously injured.

1.41 These findings suggest that, to minimise risks of accidents for pedestrians and cyclists, the road layout should be designed to exclude or discourage non-access traffic, reduce vehicle flows and restrain vehicle speeds. Section 2 gives guidance on these matters. The location of accidents in the immediate vicinity of the development site should be identified and, if necessary, the possibility should be explored of designing the layout to help reduce accident risks at these locations.

Safety on shared surface roads

1.42 The first edition of this bulletin recognised the part that shared surface roads can play in creating formal and informal landscape settings to suit different kinds of urban, suburban and rural sites, but the bulletin recommended caution until results from appraisals of these roads in use could be made available. Findings from several studies have since become available.

1.43 Appraisals of schemes with shared surface roads designed shortly before and immediately after the first edition of this bulletin was published found that the intimate scale and attractive landscape character of the surroundings provided were highly regarded by residents.[21] These roads were found to be safe and convenient for pedestrians and drivers.

1.44 A subsequent study of local accident records [22] for shared surface culs-de-sac in schemes that had been designed after the first edition of this bulletin was published, and for culs-de-sac roads with footways in a larger sample of earlier developments, found that no accidents at all had been reported on the shared surface roads. This study suggested that the use of shared surfaces will not produce any increase in reported injury accidents.

1.45 Later appraisals of other schemes [23] found that the great majority of residents who lived along shared surface roads appreciated the visual character of their surroundings and did not consider safety for pedestrians to be a problem. Specific problems such as cars travelling too fast, children playing in the road, the way cars park and difficulties in seeing cars coming were experienced mainly by residents who lived along shared surfaces that had not been designed in accordance with recommendations in the first edition of this bulletin, and where the character of the surroundings created was not significantly different from that provided along roads with footways. These results suggest that safety along shared surface roads can be ensured by good design. Section 2 contains new guidance on the design of such roads.

1.46 Highway authorities may wish to take this evidence and the guidance in Section 2 into account when considering their duty under Section 66 of the Highways Act 1980 to provide '...... a proper and sufficient footway as part of the highway in any case where they consider the provision thereof necessary or desirable for the safety or accommodation of pedestrians ..... '.

Traffic nuisance, energy use and pollution

1.47 The level of noise, fumes, vibration and other nuisances caused by vehicular traffic depends on such factors as traffic volume, the numbers and types of heavy vehicles, average speed, gradients and smoothness of traffic flow.[24] Nuisance from such factors is normally generated along distributor roads by through traffic and is seldom generated to a persistently unacceptable extent by residents in the vicinity of their homes. Nevertheless, the need to minimise risks of traffic nuisance should be taken into account when planning the layout of residential roads - for instance, by using configurations that keep vehicle flows and speeds to a minimum and that do not require service vehicles to reverse in order to turn.

1.48 The provision of convenient and attractive routes for pedestrians and cyclists and readily accessible bus services may encourage some residents to leave their cars at home for some journeys - thereby contributing to the conservation of fuel and reductions in air pollution.[25] In large developments, the road layout should be designed to keep to a minimum the distances that have to be travelled by car within the layout to reach destinations outside the site. Also, direct routes between different parts of the site should be provided for service vehicles when making collections and deliveries.

The local area

1.49 Roads and footpaths in the local area may need to be modified to help create suitable conditions within a new residential development - especially in areas where new-build and improvement schemes are intermixed. In particular, it may be necessary to consider whether distributor roads could be improved in order to discourage the use of the new residential roads by non-access traffic or to provide safer routes to local amenities for the new residents - for instance, by providing new pedestrian crossings or islands, creating new or improved facilities for cyclists, preventing obstructions to visibility caused by on-street parking or introducing measures to restrain vehicle speeds.

1.50 Equally, new development can often contribute to the creation of better surroundings in the existing area or in a proposed future development. It may be possible for instance to help exclude unsuitable traffic from existing roads, provide alternative access and off-street parking facilities for existing dwellings on distributor roads or provide safer and more convenient pedestrian or cyclist links to local community facilities.

Security from crime and vandalism

Assessing risks

1.51 The relative priority that needs to be attached to security in the design of individual developments may vary according to the likely frequency and seriousness of different types of crime and vandalism in the local area, and residents' perceptions of the risks of crimes occurring.

1.52 Risks need to be jointly assessed at the outset of design by developers, their designers and the local crime prevention officer. For blocks of flats and for rented housing it may also be important to consider the role of estate management in reducing risks and sustaining security.[26]

1.53 Whilst recognising that the causes of crime and vandalism are complex, many studies suggest that housing layout design can play a part in minimising risks - complementing any physical measures needed to strengthen dwellings against intruders.[27]

Natural surveillance

1.54 A number of design principles are now beginning to emerge. The most common thread in the evidence is the importance of designing layouts which will provide natural surveillance and some control over access, and enhance the perceived ownership of an area by its residents. The design should encourage the residents of an area to adopt it as their own and to exercise a proprietorial interest in it, so that they will feel a collective concern for its well-being. These aspects of design may act as a deterrent to potential offenders by increasing the chance of their being observed and recognised as strangers, and lead to the possibility of intervention by residents. Section 2 gives guidance on natural surveillance in relation to the layout of roads and footpaths and provision for parking.

Communal open space

1.55 The provision of large areas of communal open space may be acceptable for developments in neighbourhoods with low crime rates and where adequate arrangements for maintenance can be made (see Paragraph 1.12). But communal open space is likely to present problems where crime rates are high - because of risks of attacks, vandalism and anti-social behaviour. Where existing communal open space is adjacent to or part of a development and is thought by local residents to create problems, alternative uses may need to be considered (e.g. private gardens or allotments).

1.56 High fences and gates may be needed to deter intruders when back gardens abut communal open space or footpaths. Similar protection may also be needed at side entrances to houses. Shrubs will normally need to be planted to soften the visual impact of extensive rear garden fences, and special efforts may be needed to establish and maintain such planting where there are risks of vandalism.

The site in the urban road system

A hierarchical structure.

1.57 Local authorities are encouraged to work towards a hierarchical structure for their roads, the hierarchy being based on the roads' intended functions. This hierarchy defines primary, district and local distributor roads and residential access roads. New development and redevelopment should be designed to fit into and strengthen this hierarchy. The aim of the hierarchical approach is to influence traffic distribution in order to:

- help traffic use the main roads safely;
- discourage the use of local residential roads for through travel;
- create safe conditions for all users of residential roads, and especially young pedestrians.

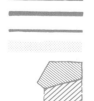

Primary distributor

District distributor

Local distributor
Existing housing, residential roads are not shown

New housing sites

Future development

30

1.58 Figure 30 illustrates a hierarchical network configuration of distributors and residential roads in an urban road system and shows sites of different shapes and sizes in different places. This diagram shows that the impact of a development on its surroundings and the functions

15

of roads within a scheme may depend upon the size and location of the site in relation to the distributor roads. For example, while sites 1 and 2 both gain access from existing residential roads, each will make a different impact upon the adjacent roads. Similarly, while sites 3, 4, 5, and 6 all gain access from local distributors, each presents different site layout opportunities and different problems regarding the exclusion of non-access traffic. Also, a local distributor is an integral part of site 6, while sites 7, 8 and 9 each gain access from district distributors rather than, as is preferable, from other residential roads or local distributors.

1.59 Subject to considerations of traffic volume, vehicle access to community facilities may be from residential roads. But where large service vehicles are likely to require access to facilities such as shops, it will normally be more sensible for reasons of economy and general amenity to provide separate road access from the local distributor.

1.60 In many existing urban areas a road hierarchy may not be easily recognisable. However, it is essential for the local authority to define the intended functions of roads in the local area so that decisions can be made about the functions of roads within the site. When assessing the functions of roads in the local area it is important to consider not just vehicular movement but the volumes and patterns of movements by pedestrians and cyclists as well.

Movement patterns

1.61 When estimating directions, amounts and types of movement it is necessary to consider what changes are likely to occur over time in the location of local facilities and in the volumes and types of vehicular traffic generated within the scheme and different parts of the road layout in the local area. While it is seldom possible to predict such changes with any certainty, those which are foreseeable, and especially those which may affect the feasibility of future development, must be taken into account when planning the layout and also when considering what impact the scheme itself will have on the surrounding area.

1.62 The volume of vehicular traffic within a scheme will depend on such variables as household size and composition, socio-economic status, levels of car ownership and use of public transport. It is important to recognise that these factors change over time, and will vary from location to location. Special considerations will apply where community facilities such as schools and shops are served by residential roads. Ideally such factors should form part of the local area planning and transportation data and it is not envisaged that exhaustive calculations would be required for each individual housing project.

20mph speed limit zones

1.63 To help reduce the number of road accidents and the severity of injuries, the Department of Transport (which has to approve any speed limit below 30mph) will approve 20mph limits in residential areas - provided drivers are alerted at the entrances to such areas that they are entering a 20mph zone and engineering measures are used to enforce the lower speeds required.[28] The Department of Transport's guidelines on such matters will need to be consulted by the local highway authority when a new residential development is located in a 20mph zone or is considered eligible for designation as a 20mph zone.

1.64 The roads' intended functions need to be taken into account when considering the designation of 20mph zones. Such zones may contribute to the creation of a hierarchical structure for the local road system and thereby help to discourage non-access traffic from using residential roads. Section 2 gives guidance on the design measures that need to be used to restrain vehicle speeds.

2

The Layout Overall

Main objectives

2.01 To help create a safe and nuisance-free environment the road layout should be designed to ensure that:

(a) non-access vehicular traffic finds distributor roads more convenient to use than residential roads;

(b) vehicle flows are low in the immediate vicinity of homes;

(c) vehicle speeds are restrained along residential roads.

2.02 The layout configuration required to meet these objectives should be designed to ensure that:

(a) residents' needs for fire, ambulance and other emergency services can be met promptly;

(b) the shortest practicable vehicular routes are provided between dwellings and points of access to the site;

(c) the most direct practicable routes between dwellings are provided for those who make regular door-to-door collections and deliveries (e.g. refuse collection and milk delivery);

(d) direct vehicular access to as many dwellings as possible is available for taxi drivers, mini-bus operators and others who provide door-to-door personal transport services;

(e) residents' needs for conveniently located public transport facilities can be met;

(f) the layout configuration, street names, signs and dwelling numbering together help strangers to find their way around.

2.03 The measures that are used to restrain vehicle speeds should be located and designed to:

(a) restrain the speeds of all vehicles;

(b) ensure that pedestrians, cyclists and drivers and their passengers are not faced with:

(i) unexpected conditions that could constitute safety hazards;

(ii) unnecessary discomfort;

(iii) avoidable inconvenience;

(c) minimise risks of traffic nuisance from increased acceleration, braking and changing gear, exhaust fumes or vibration;

(d) ensure that the great majority of drivers find acceptable the distances over which they are expected to proceed at low speeds.

2.04 Inter-visibility should be provided to suit the vehicle speeds that are likely to occur as a result of introducing measures to restrain vehicle speeds.

2.05 To provide safe, secure and convenient routes for pedestrians and cyclists the layout of roads and footpaths should ensure that:

(a) drivers are aware on entry and throughout the layout that they are in surroundings where the needs of pedestrians and cyclists take precedence over the free flow of vehicles;

17

(b) main routes between the homes, bus stops and community facilities:

    (i)   are as short as possible;

    (ii)   have the easiest practicable gradients - especially for elderly and disabled people;

    (iii) are as protected as possible from driving rain, wind and snow;

    (iv) are well-lit after dark and self-policing by being overlooked by dwellings or passing traffic;

(c) footways are provided where a shared surface would not be appropriate;

(d) shared surfaces are designed to allow pedestrians, cyclists and drivers to mix safely.

2.06 The road and footpath layout also needs to ensure that:

(a) suitable routes are provided for gas, water, electricity, telecommunication and sewerage services underground and adequate space is made available for above-ground equipment such as telephone kiosks and sub-stations;

(b) adequate street lighting is provided in all parts of the layout to enhance safety and security for drivers, pedestrians and cyclists.

2.07 To help avoid the danger, nuisance, inconvenience and damage that can be caused by indiscriminate on-street parking, it is necessary in design to provide:

(a) sufficient numbers of off-street parking spaces for residents' and visitors' cars;

(b) spaces for short-term parking by service vehicles and casual callers on or alongside carriageways which give direct access to dwellings;

(c) routes between parking spaces and dwelling entrances or other destinations that are shorter and more convenient to use than would be the case if parking occurred on carriageways;

(d) parking spaces close to and within sight of the dwellings they are intended to serve.

2.08 Verges, footpath routes and parking areas should be designed to accommodate trees and shrubs when this would help to enhance the visual character of the development or provide other benefits such as privacy and shelter from wind.

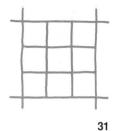

31

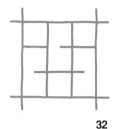

32

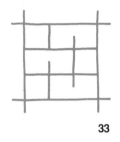

33

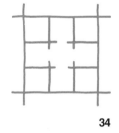

34

Non-access vehicular traffic and vehicle flows

Through routes

2.09 The changes in carriageway alignment that are needed to restrain vehicle speeds along residential roads (see Paragraphs 2.29-2.60) will often be sufficient in themselves to deter non-access traffic. In these circumstances, the most convenient practicable vehicular routes may be provided for residents and visitors - by using a network configuration of roads such as that illustrated in Figure 31.* It aims to provide frequent and conveniently distributed access points and allow service vehicles to move freely between different parts of the site. Safety and ease of traffic flow will need to be taken into account when determining the number and location of access points along distributor and major access roads serving such layouts (see Section 3).

2.10 Additional measures would be required if non-access traffic would not be sufficiently deterred by the speed restraints - for instance because of traffic delays along roads outside the site. In these circumstances, vehicular routes within the site may need to be made significantly longer than the alternatives outside. Or some routes across the site may need to be made more tortuous than those outside (Figures 32 and 33). In extreme circumstances, a mixture of loop roads and culs-de-sac may be needed to exclude non-access traffic from the site - either altogether (Figure 34) or in some directions only (Figures 35 and 36). The use of one-way streets for these purposes will not normally be appropriate for residential areas. The provision of links which could only be used by buses may sometimes need to be considered in the circumstances illustrated in Figures 34-36. The design of such links would need to be carefully considered by the highway authority in consultation with public transport operators.

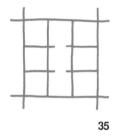

35

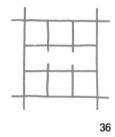

36

Vehicle flows

2.11 It will normally be possible to keep maximum vehicle flows to a minimum and distribute vehicle flows evenly in the layout by considering carefully the configuration of the road layout in relation to the number and location of access points to the site, the density of building development in different parts of the layout and the location of external attractions such as centres of shopping, schools and principal traffic routes.

2.12 The means used to restrain vehicle speeds along residential roads will normally encourage residents and visitors to take the shortest practicable routes when entering and leaving the development. Therefore, vehicle flows would tend to be well-distributed in the layouts illustrated in Figures 31-36 and maximum vehicle flows would largely depend upon the numbers of access points provided. For example, with the eight access points illustrated in Figure 31, maximum flows

*This and other layouts in this section are intended to illustrate an individual aspect of design in a simplified and diagrammatic manner. They are not intended to be examples of realistic layouts.

would be no more than around 40 vehicles per hour if there were 320 dwellings in the development (assuming one vehicle journey per dwelling in the peak hour), similar building densities in different parts of the area and well-distributed external attractions. But such flows would increase if the numbers of access points were reduced. For instance, they would be up to around 80 vph with only four accesses (Figures 37 and 38), 160 vph with only two accesses (Figures 39 and 40) and 320 vph with only one access (Figure 41). Vehicle flows would tend to be divided unequally between access points and distributed unevenly in different parts of the road layout if asymmetrical layout configurations and arrangements of access points were used (Figure 42).

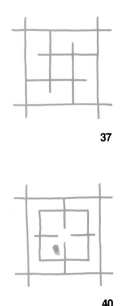

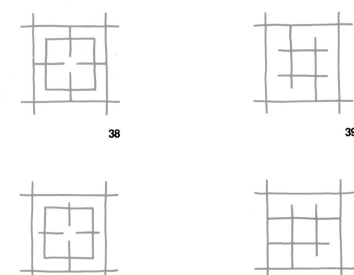

37 38 39

40 41 42

Access to dwellings

Direct access

2.13 For economy, every effort should be made in design to ensure that most stretches of road in the layout directly serve dwellings. As general guidance, it is suggested that access may normally be provided from residential roads serving up to around 300 dwellings. Ways to ensure that most residential roads serve well under 300 dwellings have been suggested above.

2.14 There may be some places in large developments where vehicle flows are high enough to warrant the avoidance of direct vehicular access to dwellings. Such places may occur where several large groups of dwellings have to be connected to the distributor road by a short length of road and a single junction (Figure 41). These are sometimes called 'transitional' roads. They always form the stem of a T-junction with a distributor road. Such roads should be kept as short as practicable.

Visual character

2.15 To help enhance the visual character of the development, the layout as a whole should be designed so that the fronts of dwellings (not gable walls, fences or garages at the ends of back gardens) face most stretches of road. Special efforts will be needed to ensure that houses face onto local distributor and other roads that serve more than around 300 dwellings (and which consequently do not provide direct access

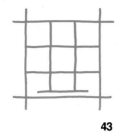

43

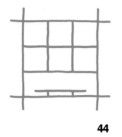

44

to dwellings). For example, the residential road layout may need to be extended to provide an access road (or shared driveway) adjacent to and running in parallel with the major road (Figure 43).

2.16 Alternatively, the access road may be connected to the major road (Figure 44) provided:

(a) the spacing of connections along the major road is not less than the minima required for junctions;

(b) most drivers do not normally have to park on the major road* (i.e. assigned and unassigned parking provision for both residents and visitors is adequate);

(c) drivers of cars and small service vehicles would not have to reverse out onto the major road (i.e. the design of the access road and individual driveways together allow such vehicles to turn).

2.17 To help enhance the appearance of these layout arrangements, trees should normally be planted between the access road and the major road. Tree and shrub planting is essential where there would otherwise be unrelieved and extensive views of garden fences and screen walls.

Natural surveillance

2.18 The location of dwellings to face most stretches of roads will also help to enhance security - provided the dwellings are designed in detail so that natural surveillance is possible from their windows. In this way, the layouts illustrated in Figures 43 and 44 would help to enhance security along the major road. Special efforts to provide natural surveillance may also be needed along other roads.

2.19 Either culs-de-sac, loops or through roads may be designed to enhance security by encouraging natural surveillance. Culs-de-sac may be effective if they serve only small numbers of houses and are not linked together by footpaths, and if the houses are grouped to overlook each other across the road or a small common open space. Equally, culs-de-sac, loops or through roads serving larger numbers of houses may also be effective - with the greater volumes of passing traffic providing surveillance. However, the rate of traffic flow may be crucial - too little and surveillance may be reduced, too great and residents may turn their back on the street.

Alternative means of access

2.20 The road layout should be designed to help minimise risks of access problems for the emergency services and residents arising as a result of vehicle breakdown, road maintenance or the need to gain access to services underground. While none of these events are likely to occur very frequently they may block access or cause inconvenience unless means can be found to by-pass them.

2.21 Risks of such problems arising can mainly be reduced by providing alternative means of vehicular access - either permanently or for use in an emergency. This should be the aim wherever the road layout serves a significant number of dwellings. When considering the relative merits of these design options it should be borne in mind that:

(a) either permanent or emergency alternative means of access can help to provide direct and convenient routes for pedestrians and cyclists;

---

* Subject to the highway authority's views about the intended function of the road and its suitability for occasional parking by refuse and other service vehicles that may choose not to enter the access road.

(b) the use of loops and through roads to provide permanent alternative means of access will also - compared with culs-de-sac:

   (i) reduce the nuisance and inconvenience that can be caused by vehicles reversing and turning,

   (ii) distribute vehicle flows more evenly over the layout.

2.22 As general guidance, it is suggested that:

(a) a road serving up to around 50 dwellings may be either a loop or through road, a cul-de-sac with a footpath link that could be made available for use by vehicles in an emergency or a cul-de-sac without such a footpath link;

(b) a road serving more than around 50 dwellings and up to around 100 dwellings should preferably be a loop (Figure 45) or through road, or at least have a footpath link for use by vehicles in an emergency (Figure 46);

(c) for a road serving more than around 100 dwellings and up to around 300 dwellings

   (i) two points of access should be provided to the part of the site being served and the road layout should conveniently connect those points of access (Figure 47) or, where only one point of access is available,

   (ii) the road layout should form a circuit and there should be the shortest practicable connection between this circuit and the point of access (Figure 48).

2.23 Special measures to maintain access may need to be considered when alternative access cannot be provided in the manner described above and a road serves more than around 300 dwellings.

Access for service vehicles

2.24 The convenience of those who make regular collections and deliveries and provide door-to-door personal transport services will be affected by the road layout configuration. This should be taken into account when planning the layout. For example, the layouts in Figures 49-52 all have the same travel distance from the entry point of the road to the furthest dwelling. Figures 50-52 however, could, with the same dwelling frontages, serve twice as many dwellings as Figure 49. But in Figures 50 and 51 this could only be achieved at the expense of doubling the traffic volume at the entry point and in Figure 52 care would be needed to ensure that the flows at each entry/exit point were roughly equal. Figures 51 and 52, however, have an advantage over Figures 49 and 50 in halving the distance travelled by service vehicles such as dustcarts and milk floats for the same number of dwellings.

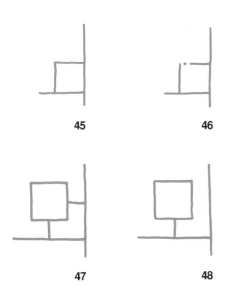

45   **46**

47   **48**

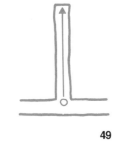

49

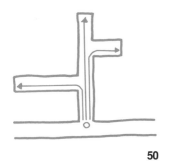

50

51

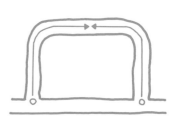

52

2.25 Local public transport operators' requirements will need to be taken into account at the outset of design when buses are likely to use the road layout.[29] In these circumstances, bus stops should be located and designed to ensure that:

(a) residents will find them convenient and safe to use (usually near road junctions, grouped with telephone kiosks and post boxes, with shelter provided from wind and rain, with good street lighting and in places which can be seen from dwelling windows and by passers-by);

(b) bus drivers will find them convenient to use (usually at straight stretches of kerb - not lay-bys - and in places where parked cars will not prevent the buses from approaching and departing in a straight line);

(c) when necessary, adequate space is provided for buses to turn or to wait (dimensions and configurations should be determined in consultation with the local bus operators);

(d) the least possible nuisance is created for residents living in nearby houses;

(e) the overall visual character of the road is enhanced (for instance, by carefully selecting and designing the signs and shelters, by using any large space required for buses to turn to plant large trees and by integrating all of these features into the overall design concept for the road).

Vehicle speeds

Driving speeds

2.26 Regardless of statutory speed limits in built-up areas, all drivers are legally obliged to drive with caution to suit the prevailing conditions. These conditions are related to road layout, matters of detailed design such as the alignment and widths of roads, and to the patterns of vehicular movement and pedestrian use to be expected on and around the roads. It has to be recognised, however, that many drivers exceed statutory speed limits - if the road alignment allows them to do so - and some drivers proceed at speeds exceeding those which suit the prevailing conditions. These considerations need to be taken into account when designing the roads and their surroundings to restrain vehicle speeds.

Visibility

2.27 Restricted visibility in the absence of other precautions cannot be considered a safe means of reducing vehicle speeds. For safety, drivers must be able to see a potential hazard in time to slow down or stop comfortably before reaching it. It is necessary therefore to consider the driver's line of vision, in both the vertical and horizontal planes, and the stopping distance of the vehicle.

2.28 Visibility distances must be adequate for the expected speed of vehicles. If measures have been taken to keep the speed of most vehicles below 30mph it may be possible to base visibility on these lower speeds. In such cases, likely vehicle speeds along each stretch of carriageway and at each junction and bend should be considered separately - together with the location of any potential obstructions to visibility such as buildings, planting and summits. Section 3 gives detailed guidance on visibility considerations in general and particular requirements for visibility at junctions, on bends and along the edges of carriageways.

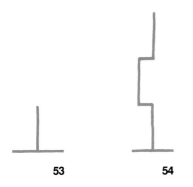

53   **54**

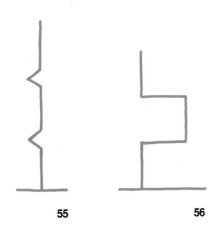

55   **56**

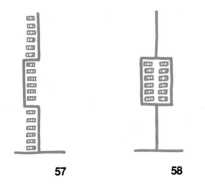

57   **58**

2.29 Section 62 of the Highways Act provides a general power to improve highways; Section 75 of the same Act empowers a highway authority to vary the relative widths of the carriageway and any footway; and Section 77 of the Highways Act 1980 permits the level of a highway to be raised, lowered or otherwise altered. Specific powers relating to road humps, including the necessity to post and publish notices, are contained in Sections 90A to 90F of the Highways Act 1980.[30] Highway authorities will need to be satisfied that any speed restraint measures constructed come within the meaning of the relevant legislation. The Department of Transport's guidelines for the introduction of 20mph speed limits may also be relevant (see Paragraphs 1.63 and 1.64).

2.30 The adequacy of speed restraints in this respect and their acceptability to different road users will depend upon the types of restraints used, their spacing and detailed design, the combined impact of different types of restraints, the effectiveness of any complementary design measures and the numbers and types of restraints that have to be negotiated along the road in question. It is important that the locations of speed restraints should be clearly apparent to drivers, cyclists and pedestrians during the day and at night.

2.31 Experience of designing speed restraints for new developments in this country is limited so far. And evidence from appraisals of restraints in use is even more limited. Consequently it is only possible in this edition of the bulletin to make some general observations about the main considerations that need to be taken into account in design and to describe some of the design options that are currently available.

2.32 Local authorities may wish to encourage the use of some types of speed restraints on a small scale in the first instance, appraise their benefits and drawbacks and then produce local design guidance in light of this experience. Experience gained from introducing speed restraints along existing roads may also provide relevant information.

Changes in horizontal alignment

2.33 An indication of the likely consequences for vehicle speeds of introducing significant changes in horizontal alignment is provided by findings from a study of residential streets with traditional cross-sections and visibility standards.[31]

2.34 The study found a clear relationship between vehicle speeds and the length of the street between junctions or bends. For straight lengths of 60m and 100m the findings suggest that the 85 percentile speeds were close to 20mph and 25mph respectively. Over 200m, such speeds were well in excess of 30mph.

2.35 Vehicle speeds in a small sample of very long culs-de-sac and loop roads were lower than those found along other roads with comparable lengths. Higher speeds were found along access roads which functioned as and had the appearance of distributors.

2.36 The study also found that vehicle speeds were related to radii at bends, but the effects of tight radii were not as great as had been expected - probably because of drivers cutting corners.

2.37 These findings suggest that vehicle speeds in new residential developments may be effectively restrained by using short culs-de-sac, carriageway offsets and junctions or small radius 90 degree bends to keep unrestricted lengths of roads to a minimum (Figures 53-56). Also, grouped parking spaces may be used in conjunction with carriageway offsets to break up what would otherwise be excessively long stretches of road (Figures 57 and 58).

2.38 Mountable shoulders have been used in some developments to help reduce the speeds of cars at bends. The bends have two different radii - to suit the turning characteristics of cars and larger vehicles - and the shoulder is made uncomfortable for car drivers and, to a lesser extent, for the drivers of larger vehicles. This speed restraint has been used on bends along shared surface roads (Figure 59) and access roads (roads with footways) (Figure 60).

2.39 Mountable shoulders have also been used to help reduce the speeds of cars turning at junctions between two shared surface roads (Figure 61). This restraint should not be used for junctions where roads have footways, because the shoulder may divert pedestrians crossing the mouth of the stem road onto the carriageway of the through road.

59

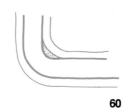

60

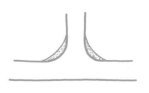

61

2.40 Chicanes and islands (Figures 62 and 63) are two other changes in horizontal alignment that have been used to restrain speeds. Their effect on drivers is similar to that of having to manoeuvre around parked cars.

2.41 To be effective, the lateral displacement of the running lane in a chicane must be severe, and the length of the displacement should be short. Similar care is needed in the design of islands. For instance, mountable shoulders should normally be used to restrain the driving speeds of cars and the central part of the island should be designed to prevent service vehicles driving over it.

62

2.42 Adequate reductions in vehicle speeds will be feasible in the great majority of new developments through the use of short culs-de-sac and changes in horizontal alignment alone. These restraints cause the least possible discomfort and inconvenience to cyclists, drivers and their passengers and to pedestrians using shared surface roads. Also, they are familiar in one form or another to all road users today. For these reasons their use will be appropriate for most new developments.

Changes in vertical alignment

2.43 However, it needs to be recognised that site constraints or the overall design concept for a development may make changes in vertical alignment preferable - either in some parts of the layout or as the main measure used to restrain vehicle speeds. In large developments, changes in vertical alignment may be needed to avoid creating the maze-like layout configurations which can result from using only changes in horizontal alignment.

63

2.44 Road humps conforming to the Highways (Road Hump) Regulations 1990 can be appropriate for both new developments and improvement schemes.[32] The flexibility conferred by the 1990 Regulations permits the construction of road humps to form speed tables and raised junctions (i.e. the use of ramps to raise the carriageway level to create a plateau). These are the changes in vertical alignment that seem

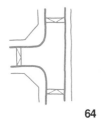

64

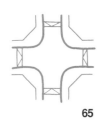

65

66

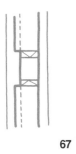

67

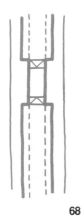

68

most likely to be acceptable to road users. Such humps must be an integral and acceptable part of the overall design concept - not an afterthought.

2.45 It should be noted that the Regulations permit humps to be used in 20mph zones without many of the warning signs and other restrictions which apply to their use on other roads (though these humps - like any other speed restraint - should be clearly visible to drivers and others during the day and at night (see Paragraphs 2.54 and 2.55).

2.46 Raised junctions may extend over the whole area of a T-junction to restrain speeds along the priority road (Figure 64) or at cross roads (see Paragraph 3.38) to restrain speeds in both directions (Figure 65). A series of speed tables may be used to interrupt what would otherwise be an excessively long stretch of road (see Paragraph 2.56) (Figure 66). Also, where space allows, sheltered parking bays may be provided along one or both sides of the carriageway and speed tables used to slow speeds in places where footway extensions are located (Figures 67 and 68). Speed tables may also be used to emphasise pedestrian crossing places - for instance where a footpath route crosses a road.

2.47 Experience in other countries suggests that raised junctions and speed tables can be as effective as conventional round topped road humps in controlling vehicle speeds.[33] However, they have not been studied extensively in this country and, until this gap can be filled, it will be necessary for highway authorities to make estimates based on findings from British studies of round topped humps.

2.48 An indication of the likely effectiveness of changes in vertical alignment can be found from a study of round topped road humps (100mm in height) that were installed along existing residential roads.[34] The study found that, before the humps were installed, a large proportion of drivers exceeded the 30mph speed limit whereas when the humps were present less than 5 per cent of drivers did so. Speeds varied in relation to the spacing of the humps. For spacings of 60m and 100m the 85 percentile speeds were 20mph and 23mph respectively - very similar to the speeds cited in Paragraph 2.34.

2.49 The effectiveness of raised junctions and speed tables and their acceptability to road users will depend on their spacing and matters of detailed design such as the height, gradient and profile of the ramps, the length of the raised area and the types of surfacing materials used. In a broader sense, the overall appearance of these speed restraints may also affect their acceptability to road users.

2.50 When considering the use of speed tables or kerb to kerb flat top road humps it should be borne in mind that such arrangements may be hazardous for blind and partially sighted people who are unable to distinguish between the carriageway and the footway. Tactile surfaces laid in accordance with the Department of Transport's advice should be provided at these and similar locations.

Combined measures

2.51 The study described in Paragraphs 2.33-2.36 found that minor reductions in available carriageway width made little difference to speeds, and drastic ones like those produced by lines of parked cars had only a limited effect*. However, a limited number of speed measurements were also made on thirteen estates with innovative road layouts (designed before or shortly after the first edition of this bulletin

---

* Other studies have found specific carriageway narrowings to be more effective, particularly if angled or involving lateral displacement. Less severe displacements achieved by parking re-arrangements and central islands have only a small effect on mean speeds but can have a larger effect on 85th percentile speeds.[36]

was published) that included short culs-de-sac, carriageway narrowing, sharply curved roads, shared surfaces and a great variety of complementary landscape features.[35]

2.52 At nearly all points where vehicle speeds were measured, the mean speeds were well below 20mph. Though it was not possible to undertake a systematic investigation, the findings suggest that the combinations of restraints may have been effective in reducing speeds, and that the effectiveness of restraints may have been enhanced by changes in pavings, planting and other features which indicated to drivers that they were in residential surroundings where careful driving at slow speeds was expected - features such as:

(a) curving alignments and varying carriageway widths;

(b) trees, bollards and buildings forming gateways at road entrances and at narrowings and delineating changes in direction at 90 degree bends, chicanes and islands;

(c) low shrubs and hedges delineating the boundaries between carriageways and private and common open spaces (see guidance on visibility in Section 3);

(d) changes in surface materials and edge restraints highlighting the location of speed restraints and reducing the apparent widths of carriageways.

2.53 The combined visual impact of speed restraints and complementary measures must be an integral part of the overall design concept for the development.

Indicating potential hazards

2.54 Street lighting is necessary to illuminate bends, chicanes, islands, raised junctions, speed tables and road humps and the presence of these restraints should also be made conspicuous by the use of landscape features such as changes in surfacing materials, trees, shrubs or bollards. Road humps need to have warning traffic signs erected and placed in accordance with the Highways (Road Hump) Regulations 1990. However, in 20mph zones, these same regulations permit considerable relaxation as to the traffic signs that need to be erected.

2.55 Differences between carriageway surfacing materials must remain apparent and provide a safe surface for vehicular movement over the years and also be economical to maintain and remain visually acceptable after repairs have been carried out.

Target speeds and spacing of speed restraints

2.56 Driving speeds of 30mph may be justified on grounds of convenience or journey time along some local distributor and major access roads, but such roads are not provided in most residential developments. As a general guide, taking into account all the considerations set out in this section, it is suggested that the design of new residential developments should normally aim to restrain 85th percentile vehicle speeds to:

(a) well below 20mph along shared surface roads - by keeping unrestrained road lengths to no more than around 40m;

(b) about 20mph along minor access roads - by keeping unrestrained road lengths to no more than around 60m;

(c) under 30mph along major access roads - by keeping unrestrained road lengths to no more than around 80m - 120m.

2.57 On large sites the layout should provide a convenient transition between these roads (Figure 69).

2.58 Also, the design should restrain 85th percentile vehicle speeds to well below 20mph immediately outside schools and at any other points in the road layout where children may be especially at risk.

2.59 Because the road frontage occupied by speed restraints can adversely affect the provision of entrances to driveways, the location and design of these restraints should be carefully considered where roads give direct access to dwellings. The careful location of off-street and on-street parking spaces may also be critical - for, especially at bends and in places where carriageways are narrowed, indiscriminate parking may otherwise block the carriageway.

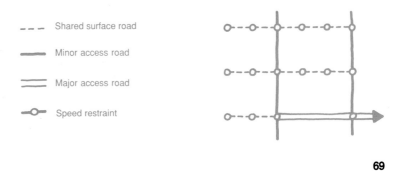

- - - Shared surface road

━━━ Minor access road

═══ Major access road

━○━ Speed restraint

69

Numbers and types of speed restraints

2.60 To minimise inconvenience for residents and those who provide essential services, every effort should be made to keep to a minimum the number of speed restraints that have to be negotiated between individual homes and roads where speeds of 30mph are acceptable - taking into account the types of restraints that are used and their configuration in relation to each other.

Direction finding

2.61 The means used to restrain vehicle speeds, exclude non-access traffic and reduce vehicle flows should help to ensure that pedestrians, cyclists and drivers who are strangers can readily find their way around (see also Paragraph 1.19).

2.62 Strangers commonly have difficulty in finding addresses when layout configurations are complex and maze-like, street name-plates are poorly sited, parking spaces are in obscure locations away from the homes, and where the main dwelling entrances and numbers cannot easily be identified.

2.63 Requirements for street names, dwelling numbering and lighting should be taken into account when planning the configuration of the layout as a whole. Dwelling numbering should be as simple as possible for visitors to understand, minimise the need for signposting and be clearly visible from the highway. It may also be necessary to circulate maps of a new area to those who provide regular and emergency services.

Provision for pedestrians and cyclists

Footpath and cycle track links

2.64 Footpath and cycle track links between roads in the layout and between the layout and the surrounding area should be created when such links would provide routes that are significantly shorter than those along the residential roads and when pedestrians and cyclists would thereby be able to reach their destinations without having to use heavily trafficked distributor roads adjacent to the site (Figure 70).

2.65 Footpath and cycle track links should be kept as short as possible - with the ends intervisible - and the layout and planting should not provide hiding places for criminals or obscure lighting. Also, whenever possible, the road, footpath and cycle track layout should enable such links to be busy and overlooked from dwellings. One layout arrangement that achieves these aims without involving the provision of footpath access to dwellings is illustrated in Figure 71.

2.66 It shows abutting turning spaces that serve the driveways of adjacent houses and allow the space from one cul-de-sac to continue into the other to create a continuous route for pedestrians and cyclists. This arrangement allows pedestrians and cyclists to take direct routes through the layout and has the added advantage of providing sufficient space to plant relatively large trees. Bollards, planting or other design measures are required to prevent vehicles passing through. Different turning space configurations may be used to create alternative layout arrangements.

2.67 Cycle gaps should be provided where turning spaces abut.[37] These gaps may also need to be designed so that they could provide through access for vehicles in an emergency. Arrangements to prevent motor cyclists passing through these gaps may need to be considered if there is likely to be a significant problem in this respect, but barriers inconvenience pedal cyclists by requiring them to dismount and the visual character of such arrangements would need to be carefully considered.

2.68 Where, space permits, cyclists are best segregated from pedestrians. However, where this is not possible, the footpath should be designed as a shared unsegregated cycle/footpath to allow cyclists to ride along it.[38] Section 3 contains relevant guidance.

Shared surface roads

2.69 The layouts described so far in this section may all be achieved with access roads (roads with footways). Shared surface roads may also be used if carefully designed (see Paragraphs 1.42-1.46).

2.70 As a general guide, it is suggested that shared surface roads may normally serve up to around 25 dwellings in a cul-de-sac and around 50 dwellings where junctions with roads with footways are located at each end of the shared surface.* These roads should be designed to provide:

(a) tight kerb radii and/or a ramp at the entrance and any changes in alignment that may be necessary to restrain driving speeds to well below 20mph (special requirements for the design of junctions with distributor roads are given in Section 3);

(b) a surface that does not give the impression of being divided into a carriageway and footways (i.e. there should be no continuous

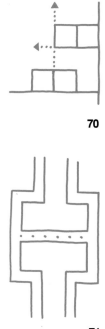

70

71

---

* Shared driveways (unadopted paved areas serving the driveways of houses) should serve no more than five houses (see Paragraph 2.81).

height differences in the cross section of the road). Spaces free from vehicular traffic may be provided (e.g. by using bollards to exclude vehicles from a play area or visibility splay). Changes in surfacing materials and colours may also be used (e.g. to help demarcate parking areas or keep vehicles away from windows and dwelling entrances or as complementary measures to help restrain vehicle speeds). Surfacing materials should be selected to suit both pedestrian and vehicular traffic in places where joint use is intended;

(c) a surface wide enough to allow pedestrians and vehicles to pass comfortably and vehicles to manoeuvre (Figure 72);

(d) clearly identifiable spaces to accommodate cars and other vehicles that need to be parked outside house curtilages (Figure 73);

72

73

74

(e) space between the shared surface and any adjacent entrances to dwellings or garages - to allow emerging pedestrians or drivers to see and be seen by approaching traffic;

(f) intervisibility between pedestrians and approaching vehicles throughout, and especially where footways, footpaths and accesses meet the shared surface;

(g) sufficient street lighting to enable road users to see potential obstacles and each other after dark;

(h) routes for services beneath or adjacent to the shared surface (see Paragraphs 2.76-2.81).

2.71 Differences between the visual character of shared surface roads and roads with footways should be further emphasised by the use of design features such as closely spaced buildings or gateways at entrances to shared surfaces (Figure 74); trees located to help demarcate parking areas or pedestrian routes (Figure 75); centre or offset drainage channels; contrasting paving materials and other elements such as ground cover shrubs planted in verges. The edge details around shared surfaces should normally be in contrasting materials and significantly lower in height than those used along access roads.

75

2.72 Planted verges may be used to discourage children from stepping off footways along the busiest major access and distributor roads or to strongly discourage drivers from parking on footways or other areas alongside carriageways. Also, when a publicly adopted and maintained verge instead of a footway is needed to make provision for services underground or to allow for vehicles to overhang, it may be possible to enhance the visual character of the development by planting appropriate shrubs in the verge. Shrub species and any proposals to make such verges wide enough to accommodate trees should be agreed with the services providers.

2.73 Sizes of trees and shrubs when mature and their location in the layout should be considered in relation to requirements for visibility, street lighting and daylight and sunlight in dwellings and gardens. Also, trees located in paved areas should not present hazards for blind and partially sighted people.

Common open space

2.74 Common open space - especially play areas - should be located near busy footways or footpaths, be clearly visible from neighbouring dwellings and be well-lit during hours of darkness. The design should not provide secluded spaces or possible hiding places. The provision of a defined footpath with good artificial lighting will encourage people walking through the area to use that route. This can help to improve the natural surveillance of the open space, and of houses backing onto it, whilst people who stray from the paths are more likely to be noticed.

2.75 Bearing in mind that requirements for casual surveillance need to be balanced with other important housing considerations such as the provision of adequate privacy and the use of trees and shrubs to enhance the visual character of the development, it is suggested that the main entrance doors, windows and side entrance gates to the houses should be visible to neighbours or passers-by.

Statutory and other services

General requirements

2.76 As well as making provision for pedestrian and vehicular movement most residential roads and footpaths provide routes for statutory and other services underground.[39] These services are an essential and integral part of the layout and their efficiency and safety in use are vital. The availability and location of existing services must be identified at the outset and those who will provide the new services must be consulted at the earliest stage of design, their requirements co-ordinated in the layout, and a balance struck between their needs and other housing objectives. The locations of existing trees and shrubs and proposals for new planting will require special consideration.

2.77 It is in the interests of residents that all services should be economical to install and maintain. The provision of adequate access for operational purposes is essential for residents in the scheme and often for users elsewhere as well.

2.78 Unnecessary capital expenditure and costs in use may be incurred if services, roads and footpaths and the location of trees and shrubs are not planned together and with care for matters of detail. In places where the most effective form of provision may be a multi-way duct or common trench for all underground services, such an option must be jointly considered at an early stage in design.

Preferred routes

2.79 Preferred routes for services are beneath publicly adopted and maintained footways, footpaths, verges or carriageways. Where services are to be laid in a publicly adopted but privately maintained verge adjacent to the carriageway:

(a) the location of the verge should be indicated by markers such as setts or bricks at private driveway or footpath crossings;

(b) the status of the verge should be made clear to purchasers in the conveyancing of the dwelling - to make them aware that:

   (i) they should not build walls or fences or plant trees or shrubs on the verge;

   (ii) services providers may need to excavate their services;

   (iii) if a cable is installed in such a verge, it could be a potential hazard to occupants who dig indiscriminately over the verge.

2.80 Where no alternative exists it may be possible to install service strips in land that is not publicly adopted providing early discussions are held with service providers and the highway authority and adequate safeguards are provided. Conveyancing documents must incorporate perpetual rights for service providers within the service strip. In addition, the provisions set out in Paragraph 2.79 (a) and (b) must be included.

Shared driveways

2.81 When unadopted shared driveways are to be used, services providers will not normally lay their mains beyond the highway boundary. Services will then be provided from the main, with the service cables/pipes laid under the surface of the shared driveway and then into each property. The acceptability of shared driveways depends upon the following considerations:

(a) no more than five dwellings should be served by the shared driveway (this will generally come within the technical limitations of servicing from a main laid in the highway boundary);

(b) where for technical reasons services cannot be supplied from a main laid in the highway, the developer must agree with the services providers on where and how their mains are to be laid (this may involve the installation of ducts, jointing pits or chambers at the developers expense).

Lighting

2.82 Lighting must be planned as an integral part of the layout of access and shared surface roads, shared driveways, footpaths and bus stops and in conjunction with the location and anticipated growth of trees. These spaces must be well-lit, not only to help people to see where they are going but also to reduce risks of night-time accidents; assist in the protection of property; discourage crime and vandalism; make residents feel secure and enhance the appearance of the scheme after dark. The standard of lighting provided should ensure that shadows are avoided in places where pedestrians would otherwise be vulnerable. Also, lighting columns and wall-mounted and other fittings normally need to be as resistant to vandalism as possible and be placed in positions which minimise risks of damage by vehicles. Detailed advice on such matters is contained in British Standard 5489 Parts 1, 3 and 9.[40]

2.83 Lighting columns and fittings also make a major impact on the appearance of the scheme during day-time and should be planned as part of the overall design concept. Highway authorities' approved ranges of columns and fittings should be chosen with this aspect in mind as well as lighting efficiency, running costs and maintenance requirements.

Provision for parking

On-street and off-street provision

2.84 Vehicles parked indiscriminately on carriageways can block access to dwellings and cause other hazards by masking road users from each other. Thus the layout should make effective provision for off-street parking. However, there will always be casual callers and service vehicle drivers who find it more convenient to park on the carriageway. Consequently, requirements for on-street parking must also be planned for in the design.

2.85 To achieve this aim successfully it is necessary to design as a whole the total space that will be used by vehicles to manoeuvre and park - taking into account the width and alignment of the carriageways, the spacing and widths of entrances to driveways, the widths and lengths of the driveways and the widths and depths of garages, car ports and hardstandings. Geometric characteristics of vehicles turning are given in Appendix 1 and some dimensions for driveways, parking bays and communal parking areas are given in Section 3.

2.86 The general considerations that planning authorities should take into account when working out their requirements for the total number of parking spaces to be provided are set out in Section 4.

Parking activities

2.87 Six main kinds of parking activities should be taken into account when considering the location of spaces and the suitability of various parking arrangements:

> Residents' cars - long-stay (e.g. overnight or during repair) and short-term (e.g. at lunch time or during car washing)
>
> Visitors' cars - long-stay (e.g. for the evening or a weekend) and short-term (e.g. to deliver a message or have lunch)
>
> Service vehicles - long-stay (e.g. furniture removals vans or builders' lorries) and short-term (e.g. refuse collection or milk delivery).

Location of parking spaces

2.88 Experience suggests that few drivers are prepared to use parking spaces more than a few metres away from their destinations, and there are increased risks of theft and vandalism when cars are parked out of sight. Consequently, the aim should be to make each small group of dwellings self-sufficient with regard to its off-street and on-street parking provision by locating:

(a) garages, car ports and hardstandings within dwelling curtilages (Figure 76);

(b) grouped garages, car ports and hardstandings immediately outside the entrances of the houses or flats they are intended to serve and within sight of kitchen or living room windows (Figure 77) - or, when there is no direct road access, at the ends of access footpaths within sight of dwellings and passers-by;

(c) casual parking spaces for visitors' cars and service vehicles as an integral part of the carriageway layout immediately outside or in proximity to the dwellings they are intended to serve.

76

77

2.89 It should be possible for a wheelchair to be pushed from vehicle setting-down points to all dwelling entrances without the need to negotiate steps. Specially reserved off-street parking spaces should be located close to all dwellings designed for occupation by wheelchair users. Other special parking arrangements for disabled people, for example restricting the use of specific parking bays to holders of official orange badges, are not recommended.

2.90 There may also be special requirements in some places for temporary parking associated with the renewal and maintenance of services - for example, outside an electricity sub-station. Those responsible for providing services must be consulted at the outset of design.

2.91 Though parked cars must inevitably interfere with intervisibility between pedestrians and drivers in places close to the entrances to private drives and in grouped parking areas, care will be needed to ensure that they do not obscure visibility on bends and at junctions.

2.92 The local fire service should be consulted about their requirements,[41] bearing in mind that adequate means of access are required to ensure that residents' lives are not placed at risk and that the safety of fire fighters is not jeopardised. Two fire appliances together with a fire car will attend most incidents, and the possibility of subsequent attendances by the police and ambulance service should be recognised.

2.93 Access points for refuse vehicles should not normally be further away than about 25m from dustbin collection points in houses and 9m from refuse storage chambers in flats.[42] Local authorities should be consulted about their requirements.

Allocation and control

78

2.94 Privately maintained and managed grouped parking spaces such as those provided to serve blocks of flats should be located between the road (or privately maintained driveway) and the building entrances, or immediately opposite. It may in addition be necessary to locate, allocate and control the spaces so that residents are given priority for those closest to their homes. Such measures may involve providing clear markings on the ground, signs or the protection of spaces with lockable parking posts (Figure 78).

2.95 It may also be necessary to provide parking deterrents such as high kerbs, low walls, closely spaced bollards, railings or deterrent surfacing to prevent parking on footpaths or planted areas.

2.96 Special problems arise where grouped hardstandings must be located further away from buildings than is required for access by service and emergency vehicles. The service access routes required in these situations will be used and sometimes blocked by residents' and visitors' cars unless formal controls such as lockable posts, gates or chain barriers are used at the entrance to the routes.

2.97 It must be emphasised, however, that all such formal methods of control are normally only feasible when the roads or driveways are unadopted and privately maintained and managed. They are also, normally, visually intrusive and difficult to enforce. It is essential therefore to make every effort in design to provide well-located parking spaces both for residents and for service vehicles.

Visual character

2.98 To enhance the visual character of large grouped parking areas, care should be taken when selecting pavings and means to demarcate individual parking spaces. The visual impact of large areas of paving may be reduced by using a variety of materials for the parking spaces, forecourts, drainage channels and running lanes in the access aisles.

2.99 Parking areas provide an important opportunity to plant trees or shrubs (especially the larger varieties) and thereby enhance the overall visual character of a development. Also, though most people are prepared to tolerate the sight of cars in residential areas, trees and shrubs can reduce their visual impact. Thus the layout as a whole should be planned to allow sufficient space for such planting - whilst at the same time allowing (for security) the cars to be seen by passers-by or from houses.

3

The Layout in Detail

Main objectives

3.01 Decisions about the layout of carriageways, bends, junctions and turning spaces should take into account the arrangements that have been made to minimise vehicle flows, reduce vehicle speeds and make provision for parking. To provide for vehicular movement within such a context and to minimise the damage and nuisance which can be caused by vehicles the widths and alignments of carriageways should be designed to take into account:

(a) the expected volumes and speeds of vehicular traffic;

(b) the frequency with which various types of vehicles need to pass each other;

(c) the provision made for off-street and on-street parking;

(d) the availability of alternative means of access to dwellings to help ensure that carriageways are not blocked by incidents such as vehicle breakdown and carriageway repairs.

3.02 The spacing and layout of junctions should be designed to take into account:

(a) the types and numbers of vehicles likely to use the junction;

(b) the directions of movement at the junction;

(c) the extent to which delays may be caused by conflicting vehicular movement at junctions with distributor roads.

3.03 The spacing and layout of turning spaces should be designed to take into account;

(a) the sizes of vehicles expected to use them;

(b) the need to avoid vehicles having to reverse over long distances;

(c) the need to prevent parking in turning bays by the careful location of adjacent off-street parking provision.

3.04 Visibility at bends, junctions and summits should be sufficient to enable drivers to stop if necessary to avoid collision. Visibility at junctions should be sufficient to enable drivers entering the priority road to do so safely and to be seen by other vehicles approaching the junction.

3.05 Footways should be designed to take into account:

(a) the functions of adjacent carriageways;

(b) requirements for statutory and other services;

(c) the amount of pedestrian movement expected;

(d) requirements of people with disabilities;

(e) the space occupied by street furniture such as lamp posts and regular obstructions such as refuse bags waiting to be collected.

3.06 Such considerations will also normally need to be taken into account in the design of footpaths. In addition, footpaths normally need to make provision for the movement of cyclists and may need to be designed to provide access to dwellings for the emergency services and allow for the movement of maintenance vehicles.

37

3.07 Verges should be designed to take into account any requirements for services underground, clearances for vehicles to overhang or provision for trees and shrubs to be planted.

3.08 The layout and dimensions of on-street parking spaces, parking bays and forecourts in grouped parking areas and parking spaces within dwelling curtilages should ensure they are convenient to use. Grouped parking bays should be demarcated to help avoid the waste of space and obstructions that can be caused by indiscriminate parking.

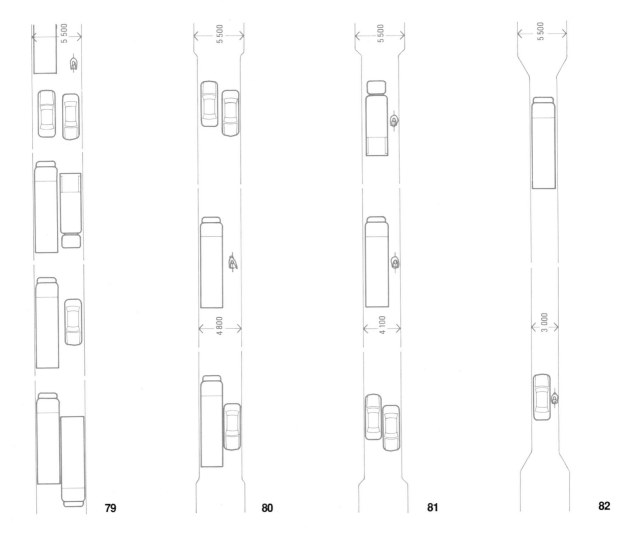

Carriageways

Vehicles

3.09 The dimensions of carriageways, junctions and turning spaces must be determined in part by the size and geometric characteristics of vehicles. Not all types of vehicles are likely to use residential roads, and those which do may not need access to all parts of a site. For such reasons a range of vehicle types and their turning and manoeuvring characteristics are given in Appendix 1. They have been chosen from studies of the types of vehicles which may generally be expected to use residential roads.* Each vehicle represents the larger size range of its type and residential roads designed to accommodate them should therefore be more than adequate for the majority of other vehicles.** The space required for vehicles to move and manoeuvre will also be influenced by the context in which they operate and the additional factors outlined below will need to be considered.

Tolerances

3.10 The ease, and hence the speed, with which vehicles may move along carriageways depends in part upon the tolerances available both between vehicles and between vehicles and kerbs. On distributor roads where ease of traffic flow is of high priority and where drivers will normally expect to be able to proceed at speeds of up to 30mph carriageway widths will need to be in accordance with the recommendations given in Roads and Traffic in Urban Areas.[45]

3.11 On residential roads however, where traffic flows are light and where journey distances are short, drivers may be expected to accept smaller tolerances consistent with the aim of restraining vehicle speeds and encouraging careful driving.

3.12 Whether or not smaller tolerances will cause unacceptable delay, reduce safety, or result in damage to footways and verges, will depend upon the types and volumes of traffic, the design of the carriageway surrounds and the distances over which drivers have to proceed. Such factors may vary considerably within a layout. The tolerances provided by various carriageway widths for the vehicles listed in Appendix 1 are shown in Figures 79-82.

3.13 As indicated in Figure 79, a width of 5.5m is normally required to enable the largest vehicles to pass each other. Most vehicles, however, will be private cars and traffic capacity will rarely be a critical issue even in larger schemes. For example a 5.5m carriageway collecting traffic from around 300 dwellings is likely to have a capacity far greater than that required, even at peak times. Such spare capacity should enable carriageways to be narrowed in places to reduce the road area and its visual impact, preserve existing features such as trees or to complement speed restraints (see Section 2).

79 *A 5.5m width allows all vehicles to pass each other, with an overall tolerance of 0.5m for the largest vehicles but with ample clearance for all others. Given the infrequency of large vehicles on residential roads, this width will normally be the maximum required to cope with residential traffic. Below 5.5m the carriageway will be too narrow for the free movement of large service vehicles such as pantechnicons. Where such vehicles are allowed access passing places may be required. The carriageway width required between passing places will then depend upon the combinations of vehicle types expected; the frequency with which vehicles may meet each other and the delay which may be caused to traffic movement. These factors may be expected to vary with traffic volume.*

80 *At 4.8m the carriageway will allow a wide car to pass a large service vehicle such as a pantechnicon with an overall tolerance of 0.5m, and traffic may therefore still be regarded as being in free flow.*

81 *At 4.1m the carriageway will be too narrow for large service vehicles such as a pantechnicons to pass vehicles other than cyclists. It does however allow wide cars to pass each other with a tolerance of 0.5m. Hence, while being more restrictive on the movement of large vehicles, a width of 4.1m will still provide two-way flow for the majority of residential traffic. Below 4m the carriageway will be too narrow for private cars comfortably to pass each other except at very low speed and may be uncomfortable for cyclists in conjunction with large vehicles. Widths of less than 4m therefore should be regarded as catering only for single-file traffic.*

82 *The choice of width below 4m will depend largely upon the frequency and ease with which cyclists and cars may need to pass each other. It is suggested that 3m be regarded as the minimum width between passing bays on a single-track system, but where narrow sections are introduced solely as pinch points a width of 2.75m will normally be adequate to allow all vehicles to pass through on their own. Widths below 2.75m may, however, be considered where private cars only are allowed access, for example in driveways. In such circumstances clearances will be necessary to allow passengers and drivers to get out of the vehicle at rest.*

---

* The vehicles illustrated in Appendix 1 do not include the largest delivery vehicle which could theoretically be used. For example, rigid vehicles can be constructed up to 12m long with a 7.5m wheelbase. Because of their size and onerous turning and manoeuvring requirements such vehicles are likely to be used primarily for inter-depot long-distance travel rather than for domestic deliveries, and have therefore not been taken into account in this bulletin. Also, Appendix 1 does not include examples of buses. Local public transport operators should be consulted about the turning and manoeuvring characteristics of the buses they propose to use. Dimensional data for buses is also available in 'Roads and Traffic in Urban Areas' and in Local Transport Note 1/91.[43]

** A possible exception to this is the fire appliance. The vehicle selected is of the type normally used for dwellings with floor levels up to 9m above ground. For medium and high-rise developments turn-table ladders and/or hydraulic platforms may be required. Such vehicles will have larger turning and manoeuvring requirements and it is essential to consult the local fire service (See also Fire Prevention Note 1/70).[44]

3.14 For the first edition of this bulletin, carriageway capacity studies were conducted at the Transport and Road Research Laboratory using single-track roads with passing places.[46] These tests indicate that with passing bays of sufficient size and frequency and with visual continuity between successive bays, undue delay or inconvenience to traffic is not caused, even where fairly high vehicle flows are carried. Requirements for visual continuity between passing places are given in Appendix 2. It is also essential to take into account any possible hazards for pedestrians or risks of blocked access as a result of on-street parking or repairs to roads and services underground. The highway authority should be consulted about their requirements regarding the use and design of single-track roads with passing places. Guidance in the remainder of this section is mainly concerned with the design of other types of residential roads.

On-street parking

3.15 Provision for off-street and on-street parking has been discussed in Section 2, and is a crucial factor when determining the width of carriageway required. Experience suggests that, wherever roads give direct access to dwellings, the carriageways are invariably used for parking by casual callers and service vehicles. Thus, as a general guide, in those locations and in any other places where parking by service vehicles will normally occur on the carriageway, a minimum carriageway width of 5.5m should be provided to allow one service vehicle to pass another that is parked. Unassigned parallel parking spaces that are contiguous with the carriageway may be within the minimum width of 5.5m - provided such spaces are located to allow a clear route and passing places for service vehicles.

3.16 A carriageway width of 5.5m will normally be sufficient to allow cars to manoeuvre around vehicles parked on the carriageway in order to get into and out of private drives. But dimensions for kerb radii, drive widths and gateways at the entrances to private drives must take into account the space that will be available for turning on the carriageway.

3.17 Along stretches of roads which do not give direct access to dwellings, the carriageway width may normally be determined by considerations of moving traffic and the availability of passing places (Figure 83). As a general guide, it is suggested that carriageway widths in places where dwellings are not directly served may normally be as follows:

	Around	Around	Up to around
No. dwellings served	50-300	25-50	25
Carriageway width (m)	5.5	4.8	4.1

Carriageway narrowings

3.18 Carriageway narrowings may be less than 4.1m, however, special attention must be paid to the needs of cyclists in these circumstances. Such narrowings will place restraints upon drivers which could cause delay and frustration and may result in actions which could jeopardize the safety of cyclists. These considerations must be taken into account when determining the extent to which the carriageway is narrowed and the distance over which it occurs.

3.19 Over long distances, such as between passing bays on a single-track system there may be a tendency for cars to attempt to pass cyclists on he narrowed section. Little is known about the safe tolerances required for these movements but, for the first edition of this bulletin, some studies were carried out by the Transport and Road

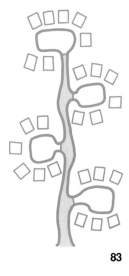

83

Research Laboratory on the reactions of cyclists to passing and being overtaken by cars on carriageways of different widths.[47] Virtually all cyclists taking part in the experiments found a carriageway width of 3.25m to be acceptable for such movements to occur, with the majority being satisfied with a width of 3m.

3.20 As general guidance, it is suggested that carriageways may normally be narrowed to 3m wide for a length of no more than about 15m to help solve a site problem (e.g. to preserve a tree) or as a complementary measure to help restrain vehicle speeds (see Section 2). Also, longer stretches may be acceptable subject to the agreement of the highway authority regarding matters such as intervisibility, the location of passing bays and provision of footways or footpaths to help ensure that pedestrians do not have to walk along the carriageway.

Emergency access

3.21 The Home Office currently recommends a minimum carriageway width of 3.66m for access roads and this advice is also incorporated in British Standard 5588: Part 1.[48] This width was specified, however, not only to enable the fire appliance to proceed at a reasonable pace but also to allow for operating space around it at the scene of the fire. Discussions with the Home Office have suggested that simply to reach the scene of the fire, carriageways as narrow as 2.75m may be acceptable to the fire service, provided that roads can be kept clear of parked cars. Fire appliances need to be able to approach to a point that is within 45 metres of a suitable entrance to any dwelling. In all cases it is essential that early discussions are held with the local fire service.

Blockages

3.22 A carriageway width of 5.5m is sufficient to allow a 3m wide lane to remain open in the event of a vehicle breakdown and it allows for patching and minor excavations to be undertaken in the remaining 2.5m wide lane. To meet other eventualities (for example to provide working space for more extensive excavations) the overall width available can normally be increased by using temporarily a part of the area that is usually occupied by the footways or verges alongside. Any services beneath would need to be temporarily protected.

3.23 Thus it should not normally be necessary to use carriageway widths greater than 5.5m for residential roads in order maintain access when vehicles breakdown or repairs to roads or services are needed. And narrower carriageway widths may be acceptable in places where adjacent footways or verges may be used to by-pass blockages. However, when they do occur, such events may cause inconvenience and damage to footways and verges and this should be taken into account in the design of those parts of the road layout where large numbers of dwellings are served.

3.24 In exceptionally busy parts of the residential road layout it may be relevant to minimise the need for excavations by locating underground services (including drainage) outside the carriageway.

3.25 Where carriageways allow only for single-file traffic in places and large numbers of dwellings are served it will normally be essential for the road layout to provide alternative means of vehicular access - either permanently or for use in an emergency (see Paragraphs 2.20-2.23).

Widening on bends

3.26 The carriageway widths required on bends involve considerations similar to those listed above but allowance must also be made for increases in the width of the paths described by vehicles when turning.

The extent of such increases will depend upon the radius of the bend and the length of the vehicles. Figure 1 in Appendix 1 shows the widths of the wheeltracks and body overhangs for various types of vehicles turning around various radii. With the criteria listed above this data may be used to determine whether and to what extent carriageway widening may be necessary.

3.27 As an example of the data applied, Figure 84 shows a carriageway width of 5.5m widened to allow for two pantechnicons to pass on a bend with a 30m outer radius. The carriageway width at its widest is made up from the swept path of the vehicle on the inside lane, the tolerances normally provided on the 5.5m straight section, and the path described by the wheels of the vehicle on the outside lane. In the latter the vehicle body has been allowed to overhang the kerbs and, while this may be permissible in situations where large vehicles are unlikely to be frequent, the body overhang must always be taken into account when determining the location and widths of footways, lighting columns, etc. Dimensions x and y are taken from Figure 1 in Appendix 1.

3.28 In this example, only the inner part of the bend has been widened. Widening the outer part of the bend would give the same effective carriageway width, but on tight bends could encourage the tendency for vehicles on the outside lane to cross over the carriageway rather than keep close to the kerb line. It should also be noted that a distance of 12m has been allowed for the realignment of each vehicle beyond its tangent point. This is taken from Figure 2 in Appendix 1 and allows for the vehicle to realign from its tightest turn. It should therefore cope with any larger radii which may be provided.

3.29 Figure 84 is intended only to illustrate how the basic data may be applied when a decision has been made about the types of vehicles which may need to pass on the bend. This may vary according to the amount of traffic on the road, and the need for widening may also be influenced by the amount of forward visibility provided between passing places on each side of the bend. On very lightly-trafficked roads, the chances of two large service vehicles such as pantechnicons or buses needing to pass on the bend may be sufficiently remote to make widening to the extent shown in Figure 84 unnecessary. Similarly where adequate forward visibility is provided between oncoming vehicles it will be possible for large vehicles to wait until the bend is clear and to use part of the opposite lane when turning. As shown in Figure 85 even with a 15m outer curve radius a pantechnicon can turn on a 5.5m carriageway without any widening and without using the whole of the carriageway width. Dimensions x and y are taken from Figure 1 in Appendix 1. The possibility of using mountable shoulders to help restrain vehicle speeds at 90 degree bends is suggested in Part 2.

3.30 In places where the bend gives direct access to dwellings it will be necessary to allow for a service vehicle to pass another that is parked on the carriageway.

3.31 As a general guide, it is suggested that carriageway widening is normally needed to the following extent on bends curving through more than 10 degrees along roads serving over 25 dwellings (widening should be on both sides of the curve, or on the inside):

Centre line radius (m)	20	30	40	50	60	80
Min. widening (m)	0.60	0.40	0.35	0.25	0.20	0.15

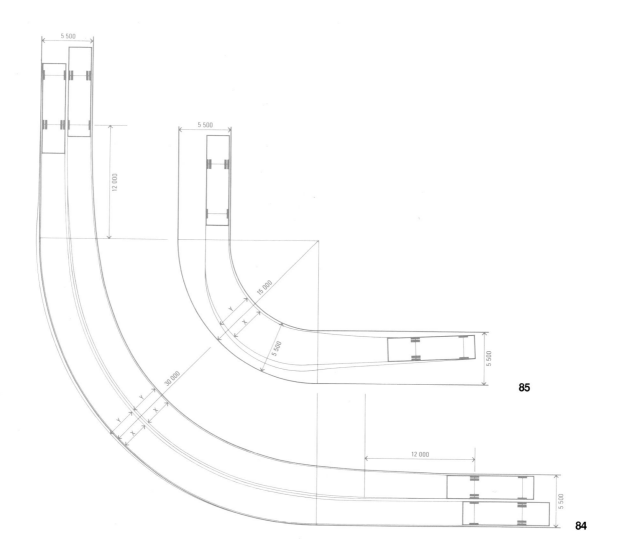

85

84

Gradients

3.32 While the need to avoid excessive cut-and-fill may often justify the use of steep gradients, regard must be paid to the difficulties which may arise in icy conditions and to their effects on vehicle speeds particularly on bends and at junctions. Steep carriageway gradients will also generally result in steep footway gradients that may cause difficulties for elderly and disabled people. On sites with significant gradients it will normally be necessary to consider carefully (and show on drawings) the relative levels of house entrances, garage floors, hardstandings, driveways, paths and adjoining roads in order to determine the practicality of the layout.

3.33 Where changes in gradients occur, vertical curves will normally be required at summits and valleys both for ease and comfort of driving and also, at summits, to ensure adequate forward visibility along the carriageway (Figures 86-88).

86   **87**   **88**

3.34 The length of curve required will depend upon the expected speed of vehicles, the angle of change between gradients and, at summits, the height from the carriageway above which clear visibility is required. For residential roads, a 600mm height is recommended (Figure 89) to provide visibility between drivers and young child pedestrians (see Paragraphs 3.56-3.59). Requirements for comfort should take into account the design measures introduced to restrain driving speeds (see Section 2).

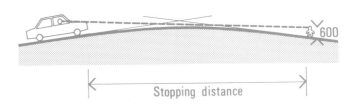

Stopping distance

89

3.35 As general guidance, it is suggested that:

(a) the gradient of the non-priority road at a junction should, whenever possible, not exceed 5% when rising towards the priority road or 4% when falling to the priority road - for a distance of at least twice the kerb radius;

(b) whenever possible, the gradient of a shared surface road should not exceed 7%.

Junctions

Traffic movements

3.36 Junctions by definition involve cross traffic and/or converging traffic movements and they are therefore potential points of hazard. There are considerations however for residential roads which do not usually apply to distributor roads and which make road layout constraints less onerous.

3.37 When the residential road layout is designed to exclude non-access traffic the vast majority of vehicle journeys will be into or out of the layout rather than between origins and destinations within the layout. The prime exceptions to this being vehicles which have to call at a number of dwellings, e.g. for refuse collection and milk delivery. The main pattern of movement around junctions may be easily predicted and will normally be less complex than on roads carrying non-access traffic. Moreover, the exclusion of non-access traffic may create a hierarchy of traffic volumes within the layout - with volumes being highest at the entry point and decreasing with each successive junction. These factors have important implications for the general configuration, spacing and detailed layouts of junctions.

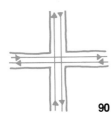

90

Configuration

3.38 Cross roads: Cross roads are generally regarded as the most dangerous form of junction, largely because they imply cross traffic movement (Figure 90). They should therefore normally be avoided. There may however be situations in which cross traffic will be virtually non-existent, for example in a short cul-de-sac layout (Figure 91), and their use should not be automatically discounted at the lower end of a road hierarchy where vehicle flows and speeds are low and where raised junctions are used to restrain vehicle speeds.

91

92

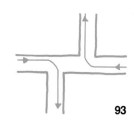

93

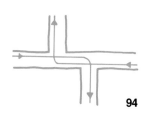

94

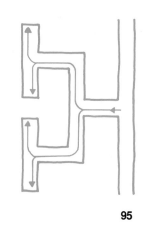

95

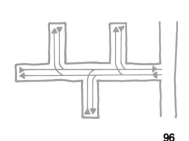

96

3.39 As the scale of the layout increases however, additional precautions may be necessary both to deal with any potential increase in cross traffic movement and to control the movement of turning traffic. The two most effective means of doing this are either to provide an island (Figure 92) or to stagger the junction (Figure 93). Mini roundabouts as described in Department Standard TD16/84 [49] and Departmental Advice Note 42/84 [50] may also be used when space permits and the requirements for signing and location suit the landscape concept for the development and allow other important housing design objectives to be met.

3.40 For staggered cross roads to work effectively they should normally be staggered by one carriageway width, and more when a large amount of cross traffic is envisaged. It is also suggested that right/left staggers should be provided in preference to left/right wherever it is likely that traffic would be turning into the side roads from either direction. This is to avoid the possibility of hooking movements (Figure 94) which can cause delay or collision.

3.41 T-junctions: Two basic forms of T-junction may be identified depending upon whether the major traffic flow is on the side (non-priority) road or the through (priority) road. The chief characteristic of the former (Figure 95) is that it automatically slows down the major traffic flow. It may therefore be used with advantage close to the entry point of the residential road layout and/or where any change in road character or traffic speed is required within the layout. Because of the direction of traffic movement determined by the layout as a whole, it is likely that the amount of cross traffic will be negligible. But there will be a corresponding increase in the amount of turning movements and it may be necessary to guard against conflicts by providing priority markings or islands where substantial volumes of traffic are expected.

3.42 Priority markings may also need to be considered for safety and convenience where the major flow is on the priority road (Figure 96) - except at the lower levels of the system where both speeds and traffic volumes can be kept low.

3.43 Y-junctions: Y-junctions are uncommon in residential road layouts largely because of the awkward shape of the land parcels which they create. Such junctions should normally have priority markings or islands except at the lower levels of the system.

3.44 As a general guide, it is suggested that non-priority roads serving more than around 100 dwellings should whenever possible join priority roads at an angle of 90 degrees, and be straight for a length of at least twice the kerb radius. Other non-priority roads may, if necessary, meet the priority road at an angle within 10 degrees of a right angle.

Spacing

3.45 The frequency of junctions along distributor roads is often an important determinant of the ease of traffic flow and of the ease with which drivers may proceed at a constant speed safely and without interruption. Generally, the closer the junction spacing, the more frequent the hold-ups to through traffic and the greater the chance of traffic build-up and accidents occurring. For such reasons the spacing of junctions along distributor roads will normally need to be in accordance with those recommended in Roads and Traffic in Urban Areas. Within the residential road layout however such considerations are seldom critical and it is important that decisions on the spacing of junctions be made with regard to the effect they have on the economic use of land.

3.46 Generally, the block spacing needed to provide adequate daylight and sunlight and to give privacy between dwellings will lead to spacings of at least 30-40m between adjacent junctions. There may be places however where closer junction spacing would enable land to be used for housing that might otherwise be wasted. It is necessary, however, to avoid the junction being blocked by traffic building up at the exit to a local distributor. The likelihood of this will depend upon the amount of traffic both on the residential road and on the local distributor, factors which may vary widely from scheme to scheme. As a general guide, it is suggested that:

(a) where the priority road serves no more than around 100 dwellings there need be no restrictions on junction spacing - and cross-roads may be used;

(b) where between around 100 and 300 dwellings are served by the priority road, it is desirable that the junction should be at least 30m (centre line spacing) from another junction on the same side of the priority road and at least 15m from another junction on the opposite side;

(c) where a residential road joins a distributor road, it should be 5.5m wide for a distance of around 20m from the junction. Footways should be provided for that distance and no junctions with other roads or accesses to driveways should normally be provided along that 20m length - to help ensure that parking does not occur on the non-priority road carriageway close to the junction.

Radii

3.47 For a given carriageway width, the radii provided at junctions will determine the ease with which a vehicle may turn and the extent to which it may do so without interfering with other traffic. Four of the potential conflict points that are related to the radii provided and which may result in delay, inconvenience or accidents are shown in Figures 97-100.

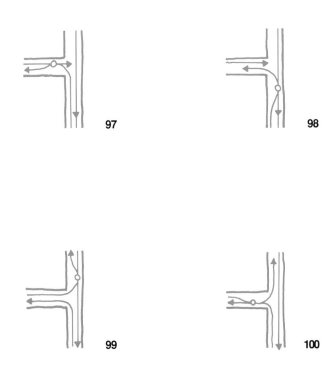

97

98

99

100

3.48 The extent to which any or all such potential conflicts need to be avoided by the choice of junction radii will largely depend upon the likely frequency and duration of the delay, which in turn will be affected by the types of vehicles involved in conflicting movements and how frequently they will meet at the junction.

3.49 The great majority of vehicles using residential road junctions and their junctions with distributor roads will be private cars, and traffic volumes will normally vary from junction to junction within the residential road layout. Requirements for residential road junctions are therefore less onerous than those for junctions between distributor roads.

3.50 The types of vehicles and movements to be catered for at any particular junction will vary according to the overall configuration of the layout. For example in Figure 101 most vehicle movements will be around on one side of the junction. Vehicles using the other corners will be mainly those needing to call at several dwellings, e.g. refuse collection and milk delivery, and it is unlikely that they will often meet other vehicles turning in the opposite direction. Figures 102 and 103 however, show two situations where all vehicles may be expected to turn in either direction. These differences may be relevant when considering the locations of features such as trees and bollards at junctions between shared surface roads.

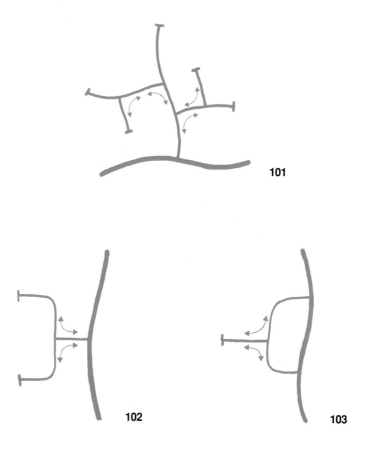

101

102

103

3.51 Most road junctions will be pedestrian crossing points and, where space permits, the aim should be to channel pedestrian movement along the priority road footway towards the tangent point of the kerb on the non-priority road and to provide dropped kerbs for prams and wheelchairs where this occurs. Elsewhere, where this is not possible, the dropped kerbs may be located on the radii. Junction radii and the space available for footways and verges may therefore have an important effect on the direction and convenience of pedestrian movement, as well as upon land use and appearance.

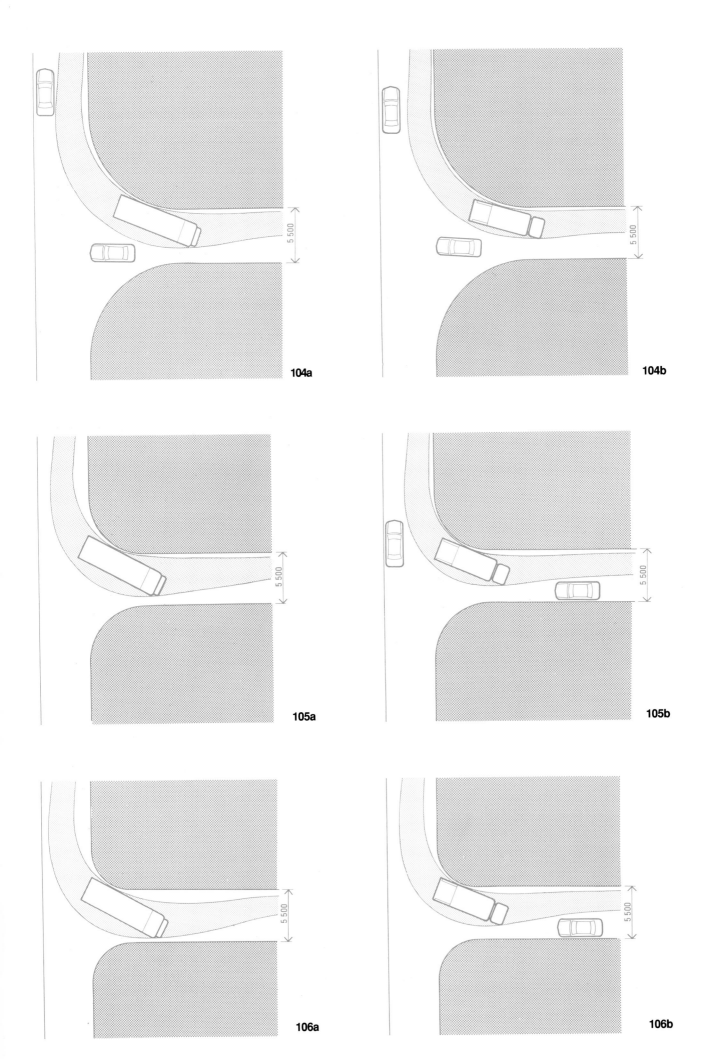

104a

104b

105a

105b

106a

106b

48

104a and b *Junctions designed to the present recommended radius of 10m for junctions with local distributor roads.[51] This radius allows both refuse vehicles and pantechnicons to turn without interfering with traffic on the priority road. Pantechnicons may have difficulty in turning past vehicles on the non-priority road and may have to wait until the junction is clear. Where the non-priority road is a residential road, however, the lightness of traffic volumes and the infrequency of vehicles of pantechnicon size mean that such delays will be very infrequent and of short duration. Radii of this order will normally be sufficient therefore for most junctions between residential and local distributor roads.*

105a and b *Junctions with 6m radii. This allows pantechnicons to turn into and out of the junction using most of the width of both carriageways, and allows refuse vehicles to turn without interfering with traffic on the priority road. Radii of this order will normally be sufficient, therefore, for junctions within the residential road layout.*

106a and b *Junctions with 4m radii. These allow all vehicles to turn into and out of the junction but require vehicles larger than private cars to use most of the width of both carriageways. Whilst this may not present problems where both roads are very lightly trafficked, the tightness of turn required for large vehicles may result in kerb mounting and precautions such as the use of bollards may be needed to prevent this happening. It is suggested therefore that radii of this order should normally be restricted to junctions carrying very low traffic volumes.*

3.52 Given the variety of circumstances which arise from the above considerations a wide range of junction forms are possible. Three examples of junction radii are shown in Figures 104-106 together with the types of movements which may occur using the turning characteristics of vehicles taken from Appendix 1. The possibility of using mountable shoulders at junctions between shared surface roads is described in Paragraph 2.39. As a general guide*, it is suggested that:

(a) kerb radii should be 10m at junctions with local distributor roads (subject to a 30mph speed limit), and 6m at junctions between residential roads where either the priority or non-priority road serves more than around 50 dwellings;

(b) 4m kerb radii may be used at junctions between residential roads where both the priority and non-priority roads are 5.5m wide and the non-priority road serves no more than around 50 dwellings;

(c) 4m kerb radii may also be used where mountable shoulders are used at junctions between shared surface carriageways - with the inner radius being 6m, or 7.5m where carriageways are narrower than 5.5m;

(d) no driveways should enter at the bellmouth of a junction.

Turning spaces

3.53 Turning spaces will normally be required whenever vehicles would otherwise have to reverse over long distances or whenever they might otherwise turn in locations which could cause damage to adjacent verges or footways.**

3.54 Depending upon the degree of inconvenience which could be caused, turning spaces may be provided as parts of junctions or else as separate elements. A wide variety of turning movements for various vehicle types are given in Appendix 1. From this data designers may construct the form and size of turning space best suited to a particular context. The diagrams give the minimum envelope required to contain the movement, but consideration must be given to the tolerances which may be necessary, how the layout of kerbs would be constructed and how the area would be swept if mechanical sweepers were used. It should also be borne in mind that turning spaces immediately in front of dwellings may be used for casual parking and this may need to be taken into account when determining the form and size of the space.

3.55 As general guidance, it is suggested that turning spaces should be designed to:

(a) allow for refuse vehicles to turn when they would otherwise have to reverse more than about 40m and for pantechnicons to turn when reversing would otherwise be for a distance greater than about 60m. When it is assumed that refuse vehicles or pantechnicons will reverse into the road:

   (i)   the road should not serve more than around 100 dwellings;

   (ii) 6m kerb radii should be provided at the junction;

(b) accommodate the vehicular movement patterns described in Appendix 1 - with space for vehicles to overhang and no obstructions more than 150mm high in these areas;

---

* The actual radii used will need to take into account any limitations associated with the selected kerb materials.

** Local public transport operators should be consulted about turning requirements for buses if any road in the layout is to serve as a bus terminus.

(c) ensure that parking spaces are provided outside the area required for turning - especially at the head of a cul-de-sac (Figure 107). The location of accesses to driveways helps to influence where drivers leave their vehicles.

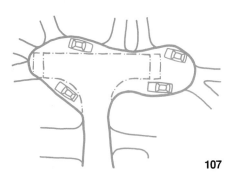

107

Visibility

Sight lines

3.56 To enable drivers to see a potential hazard in time to slow down or stop comfortably before reaching it, it is necessary to consider the drivers line of vision, in both the vertical and horizontal planes, and the stopping distance of the vehicle.

3.57 The eye level of drivers can vary from 1.05m above the carriageway in a standard car to approximately 2m in commercial vehicles. To enable drivers to see each other across summits, across bends and at junctions, unobstructed visibility will be required at least between these heights above the carriageway (Figure 108).

108

3.58 However, for drivers to see and be seen by pedestrians, particularly child pedestrians and wheelchair users*, unobstructed visibility will be required to a point closer to the ground. The height of a very young child of walking age is around 780mm, but the height of a child on a tricycle can be even lower. As general guidance, it is suggested that a height of 600mm be taken as the point above which unobstructed visibility should be provided wherever the potential exists for conflicts between motorists and young children. This will apply along most sections of residential roads and especially where shared surface roads are used.

---

*Forward visibility along carriageways will be advantageous to people who are deaf or hard of hearing, both so that they are aware of vehicle hazards and so that drivers of vehicles can see them.

50

3.59 The most obvious obstructions to visibility are summits, adjacent buildings (including bus shelters), screen walls, densely planted trees, and parked cars. Shrubs and trees may be planted in visibility splays at junctions and on bends, provided when mature they do not significantly obscure horizontal sight lines and there will continue to be clear vision between heights of 600mm and 2m above ground level. Generally the aim should be to ensure good visibility without having rely on frequent maintenance.

Stopping distances

3.60 The horizontal distance over which unobstructed visibility should be maintained will depend upon the stopping distance of vehicles. This in turn will depend upon vehicle speeds, deceleration rates and drivers reaction times. Figure 109 gives a range of stopping distances commensurate with various vehicle speeds. The distances are intended to cater for the majority of vehicles and drivers in most weather conditions and may therefore safely be used as general guidance in the design of the residential road network.

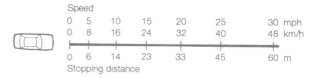

109

At junctions

3.61 To enable drivers emerging from the non-priority road to see and be seen by drivers proceeding along the priority road unobstructed visibility is needed within the shaded areas shown in Figure 110. Within the residential road network, where traffic flows are low and where the need to avoid delay has low priority, the Y dimension may be based on the expected sped of the vehicle on the priority road and hence on the stopping distance required for it to slow down or stop in order to avoid collision with vehicles emerging from the non-priority road. Stopping distance requirements along residential roads will depend upon the design measures that are used to restrain vehicle speeds and should therefore be discussed with and agreed by the highway authority.

3.62 Where the priority road is heavily trafficked however, such as on a local distributor, the objective will usually be to avoid the need for through traffic to change course, slow down or stop. In order to achieve this, drivers emerging from the non-priority road must be able to see far enough down the priority road to be able to judge when to emerge without interrupting through traffic movement. Hence the Y distance will need to be based on the time taken for vehicles to turn out from the non-priority road and the distance travelled during that time by vehicles on the priority roads proceeding at constant speed.

3.63 For the X dimension a distance of 2.4m is the minimum necessary to enable a driver who has stopped at the junction to see down the priority road without encroaching onto it. This will, however, only allow one vehicle at a time to exit safely and requires that drivers following behind should likewise stop and look. Hence, while an X distance of 2.4m may be sufficient for junctions where traffic flows on the non-priority road are likely to be low (see Paragraph 3.64c), on more heavily trafficked non-priority roads such as an exit from a large residential area or at junctions where the priority road is a major access road or a local distributor road, the distance may need to be increased to allow following vehicles to see down the priority road whilst slowly moving up to the junction point; thus allowing two or more vehicles to exit in a stream. The extent of this increase will depend largely upon the number of vehicles likely to be waiting to emerge from the junction and the extent to which delay has to be avoided. In most cases a distance of 4.5m should be sufficient for traffic volumes on the non-priority road of 300vph or less.

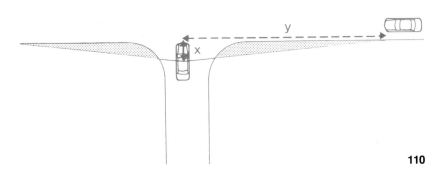

110

3.64 The following advice on visibility is set out in Planning Policy Guidance PPG 13: [52]

(a) the Y dimension (see Figure 110) will depend on the speed of traffic on the priority road: the appropriate distance can be read off Table A or Table B. If the highest traffic speed on the road in wet weather (excluding the fastest 15% of vehicles) is known* then this speed - or the next highest speed which appears on the table - should be used as the priority road speed in Table A to arrive at the appropriate Y distance. Where there is a speed limit and the actual speed of traffic on the priority road is not known it will normally be necessary to provide Y distances as shown in Table B.

Table A

Speed (mph)	75	62	53	44	37.5	30**	30***	25	20
Y (m)	295	215	160	120	90	70	60	45	33

Table B

Speed Limit (mph)	70	60	50	40	30**	30***
Y (m)	295	215	160	120	90	60

(b) the Y distances in Table A will be appropriate for priority roads where restraints have been used to reduce driving speeds to less than 30mph and about 20mph (see Section 2). The highway authority will need to be consulted about their requirements for Y distances in places where vehicle speeds are likely to be well below 20mph;

---

* Advice on measurements for this purpose is given in DTp Advice Note TA 22/81. [53]

** Where the priority road is not an access road but a higher category road.

*** Where the priority road is an access road with speeds universally below speed limit.

(c) an X distance of 9m is the normal requirement for junctions between access roads and district or local distributor roads. The provision will be required where the non-priority road is busy (e.g. where it serves as a main connection between the public road system and a housing estate development) but would not apply at junctions or accesses within estates. There, an X distance of 4.5m will normally be the acceptable minimum. For other types of access serving single dwellings or a small cul-de-sac of a half dozen dwellings, the minimum acceptable X distance is 2.4m. In urban areas with a speed limit of 30mph or less this distance may be reduced to 2m. Only in exceptional circumstances should a distance of less than 2.4m be considered for an access with multiple usage.

3.65 As general guidance, it is suggested that visibility should be ensured for vehicles turning left into a non-priority road by providing a visibility radius tangential to the kerb (i.e. inside the kerb radius). Suggested normal visibility radii (m) for different junction angles and kerb radii are as follows (Figure 111):

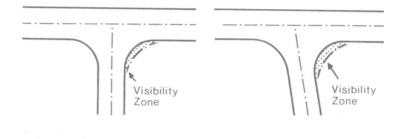

111

Junction deflection (degrees)	Kerb radius		
	4m	**6m**	**10m**
80	10m	11m	19m
90	9m	10m	19m
100	8m	9m	19m

On bends

3.66 As for junctions, the visibility required on bends should be related to the expected speed of vehicles and their stopping distances - taking into account the measures that have been used to restrain vehicle speeds. As general guidance, it is suggested that it will normally be appropriate - for simplicity - to assume that vehicle speeds will be either 30mph or 20mph - depending on which of the restraints that are recommended in Section 2 have been used and their spacing. The highway authority will need to be consulted about their requirements where vehicle speeds are likely to be well below 20mph. These speeds can be used to construct a forward visibility curve as shown in Figure 112. Forward visibility curves on bends should be constructed in accordance with the procedure set out in Figure 113:

(a) a line should be drawn parallel to the inside kerb, 1.5m into the carriageway to represent the path of the vehicle;

(b) the required stopping distance commensurate with the expected speed of the vehicle should be ascertained from Figure 109 and measured back along the vehicle path from tangent point A;

(c) the stopping distance should then be divided into equal increments of approximately 3m, and the increment points numbered in sequence;

(d) the same stopping distance with the same number of increments should then be repeated around the curve, finishing at a full stopping distance beyond the tangent point B;

(e) the area which has to be kept clear of obstruction should then be constructed by joining increments of the same number together, i.e. 1 to 1, 2 to 2 etc.

3.67 It should be noted that crossing of the carriageway may be necessary where large vehicles have to negotiate very tight bends. This may be offset by widening the carriageway but there may be situations in which it would be sensible to accept carriageway crossing. While vehicles performing this manoeuvre are likely to be travelling at low speeds, the closing speeds of opposing vehicles will need to be considered when determining the forward visibility required.

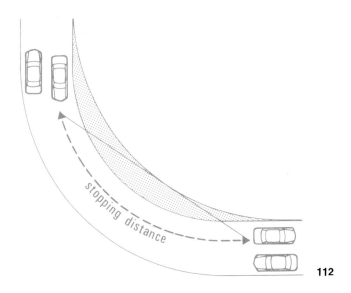

112

Along the carriageway edge

3.68 The shaded triangles defined by the X and Y distances in Figure 114 are the areas which must be kept clear of obstructions to intervisibility between pedestrians and drivers where a vehicle leaves a driveway to cross a footway. In Figure 115, where the vehicle meets the edge of a carriageway or shared surface, the X dimension allows the driver to see along the carriageway or shared surface without encroaching onto it. At the edge, conditions will be similar to that at a road junction and the Y dimension should therefore be based on the expected speed and stopping distance of vehicles on the road. As general guidance it is suggested that:

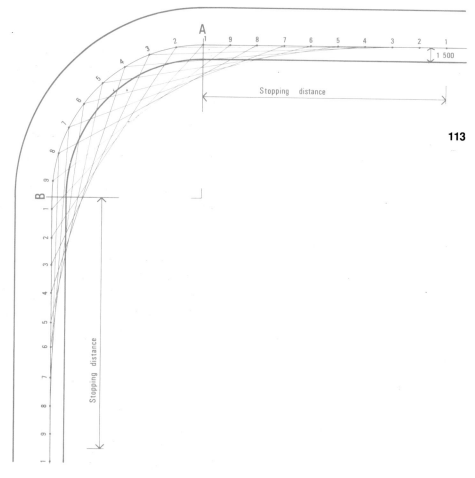

113

(a) X and Y dimensions of 2.4m should be provided where a driveway meets the back edge of a footway (Figure 114), with clear visibility at a level of 0.6m above road level in addition to visibility at the 1.05m level;

(b) an X dimension of 2.4m should be provided where a driveway meets a carriageway or shared surface (Figure 115), though in urban areas with a speed limit of 30mph or less this distance may be reduced to 2m;[54]

(c) the Y dimension where a driveway meets a carriageway (Figure 115) or shared surface should be as set out in Paragraph 3.64.

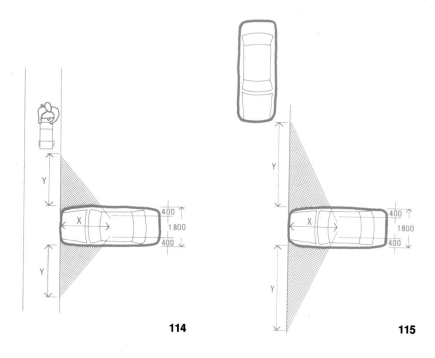

114

115

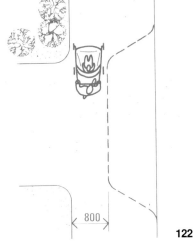

116 — 600 (750)

117 — 800 (900)

118 — 900 (1 000)

119 — 1 150 (1 250)

120 — 1 200 (1 350)

121 — 1 700 (1 800)

122 — 800

Within grouped parking areas

3.69 The absence of defined carriageways and footways within grouped parking areas means that pedestrians and vehicles normally mix in proximity both around the parked cars and on the forecourts, and it is necessary to rely for safety upon drivers proceeding at slow speed and with caution. Changes in the horizontal alignment of the access aisles may be needed to restrain driving speeds - especially in large parking areas.

Footways

Widths

3.70 The unobstructed minimum widths required for various types of pedestrian movement are shown in Figures 116-121. The dimensions shown in brackets allow for greater ease of movement.

3.71 As general guidance, it is suggested that footway widths should normally be as follows:

(a) 2m for footways along roads serving more than about 50 dwellings and where the full range of services underground are to be accommodated.* This width allows for those in wheelchairs or pushing prams to pass each other.

(b) lesser widths may be used along roads serving less than 50 dwellings* - for example, footways with a minimum width of 1.35m would allow for electric wheelchairs, allow pedestrians to pass each other and may be acceptable to services providers where the range of services is divided along each side of the carriageway;

(c) an additional footway width of 800mm (preferably in a different paving material) will be required to allow for vehicles to overhang the footway in places where vehicles park at right angles to footways (see Paragraph 3.90);

(d) where practicable, at entrances to driveways, a minimum width of 900mm carried through at footway level should be provided to enable pedestrians and wheelchair users to avoid the ramps to dropped kerbs (Figure 122);

(e) a footway width of at least 3m should normally be provided outside entrances to schools and similar community buildings;

(f) local public transport operators should be consulted about requirements for footway widths at bus stops where shelters are to be provided.

Headroom

3.72 Headroom over footways should normally be at least 2.6m - with a minimum of 2.3m for a distance no greater than 10m. Restricted headroom may extend up to a line 500mm away from the carriageway edge.

Extensions

3.73 Where on-street parking would make it difficult for pedestrians (particularly children) and drivers to see each other it may be necessary to widen the footway in places to:

(a) allow pedestrians to see beyond parked cars before stepping out onto the carriageway;

---

* Assuming that lighting columns would be located in the normal position - against the back edge of the footway. Greater widths would normally be required if lighting columns were at the front.

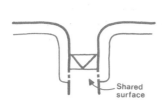

123

(b) prevent vehicles parking on the carriageways and obstructing visibility at junctions. Permanent barriers such as high kerbs or bollards may also need to be provided to prevent parking in such places.

3.74 Where an entrance ramp is provided at the entry into a shared surface non-priority road, the footways from the priority road should extend past the ramp into the shared surface road (Figure 123).

Kerbs

3.75 At road junctions and the main locations where pedestrians will be crossing residential roads, kerbs should be ramped to assist wheelchair users and those with prams or pushchairs. The gradient should be no more than 8% and the kerb should be dropped so that it is flush with the carriageway.[55] Tactile surfaces laid in accordance with the Department of Transports advice should be provided at dropped kerbs to assist blind and partially sighted people.

Footpaths

Widths

3.76 As general guidance, it is suggested that footpath widths should normally be:

(a) 2m to allow for wheelchair users and, when necessary, to allow for the full range of services underground to be accommodated;

(b) 3m as a clear corridor width for routes that are designed for occasional use by maintenance vehicles;*

(c) at least 3m outside entrances to schools and similar community buildings;

(d) 3.3m for subways and routes under buildings, or 3m for a length no greater than about 23m.[56]

Headroom

3.77 Headroom should normally be at least 2.6m - with a minimum of 2.3m for a distance no greater than about 10m.

Ramp gradients

3.78 The general aim is that ramps along footpath routes ought not to be steeper than 8%, and where possible 5%. It should be borne in mind that an extended 8% gradient may be unmanageable for some wheelchair users and pushers, and, therefore, less steep gradients are preferred. Where a gradient steeper than 8% is unavoidable, for topographical reasons, it will be a hazard for some ambulant disabled people, and, if possible, an alternative stepped approach should be available. Handrails are necessary for the safe negotiation of steps and steep ramps by disabled people.

Kerbs

3.79 At junctions between footpaths and carriageways, there should be dropped kerbs in accordance with the recommendations in Paragraph 3.75 and, when necessary, for maintenance vehicles. Visibility splays should be provided with an x dimension of 2m and a y dimension to suit the anticipated speeds of vehicles along the carriageway.

---

* Footpaths which are the only or main means of access to dwellings will need to meet the requirements of the local fire service.

Barriers

3.80 Barriers to halt child pedestrians may also be required at junctions between footpaths and carriageways. The visual character and durability of the barrier should be carefully considered. Care should be taken to ensure that guard railing does not impair driver or pedestrian visibility, particularly where children may be involved. In these circumstances, guard railing designed specifically to reduce the problem should be considered.

Provision for cyclists

3.81 Department of Transport guidance on special facilities for cyclists is available elsewhere.[57] Recommendations that may be applicable are as follows:

(a) for unsegregated use, cycle tracks may be 2m wide when combined pedestrian and cyclists peak flows will be no more than 200 per hour;

(b) for segregated use, overall width should be a minimum of 3m (1.5m each for cycle track and footpath/footway). Cyclists are best segregated from pedestrians by means of a physical separation, such as a raised kerb (with pedestrians stepping down to the cycle track). Where this is not possible, segregation by a continuous white line has been shown to be effective. For the latter situation it may be necessary to consider the needs of visually impaired people and provide tactile slabs at the start and possibly at intermediate locations so that they can be made aware of the area;

(c) appropriate signs should be provided;

(d) headroom should normally be at least 2.7m - with a minimum of 2.4m for a distance no greater than 23m;

(e) whenever possible, maximum gradients should be no more than 3% or 5% for a distance up to 100m or 7% for up to 30m;

(f) barriers may be necessary where cyclists need to be encouraged to dismount at the end of a cycle track. The minimum spacing between staggered barriers should be 1.2m to allow for electric wheelchairs;

(g) dropped kerbs should be provided at junctions between cycle tracks and carriageways, and entry into the carriageway should be at an angle of 90 degrees;

(h) visibility distances along cycle tracks (using a height of 1.05m for the line of sight of a cyclist) should be 20m on gradients less than or equal to 2%, and 26m where gradients are in excess of 2%.

Verges

Widths

3.82 Verge widths normally need to be 2m, with a minimum of 1.35m where less than the full range of services are accommodated. Section 2 describes the safeguards required where services are to be laid in publicly adopted and privately maintained verges adjacent to the carriageway and in verges that are not publicly adopted.

3.83 Where footways or verges are not provided, a 500mm wide paved margin will normally be required to provide clearance for vehicles. Where a wall or bollards abut the outer edge of the margin the carriageway surface may be extended for the 500mm distance without an upstand.

3.84 Shrubs may be planted provided they do not obscure visibility or cause harm to services beneath. To ensure good visibility without having to rely on frequent maintenance, the growth potential of shrubs planted in verges should be under 600mm in height.

3.85 Trees planted in verges should be located at least 1m away from the carriageway edge and they should not obscure sight lines when planted or as they mature.

3.86 The species of trees and shrubs selected should not cause damage to adjacent pavings, buildings or services underground - or protection should be provided.[58] Also, the trees selected should not create droppings that could cause damage to paint on cars parked beneath.

3.87 Plant selection must take into account the characteristics of the site, hardiness and commercial availability as well as size, shape, colour, growth rate and ability to withstand pruning. Long-term maintenance is a major consideration. Care should be taken to select trees and shrubs that will retain their natural character with only moderate maintenance. Also, protection may need to be provided at the base of trees to prevent damage from vehicles overhanging the verge or from mowers in grassed verges. Highways authorities and local services providers should make information available to developers about appropriate species in their areas.

Protection

3.88 Grass strips between footways and roads and small and isolated shrub beds, whilst attractive if well maintained, are often neglected and over-run by vehicles. The use of hard paving together with trees is often to be preferred in such places. When the possibility of over-running vehicles cannot be avoided, it may be necessary to either strengthen the verges or use features such as robustly constructed kerbs of a sufficient height, banks or bollards. Appropriate measures will take into account the proposed function, configuration and visual character, of the road.

Parking areas

Driveways

3.89 As general guidance, it is suggested that driveways serving garages within dwellings curtilages should normally:

(a) be long enough to accommodate a car parked in front of the garage and enable the garage door to be opened without the car having to project beyond the curtilage onto a footway (Figure 124) or shared surface (Figure 125). This length should preferably be 6m or at least 5.5m. An additional length would be required to allow for a gate at the entrance to the driveway to be opened inwards (such gates must not open out over footways or carriageways). A shorter driveway length (3m or at least 1m) may be acceptable where only small numbers of dwellings are served by a shared surface and the layout provides spaces for residents cars to be parked immediately adjacent to the drive (Figure 126) or outside the curtilages;

(b) be wide enough to allow access to both sides of the parked car and also, on one side, allow for a pathway to the house. This width should preferably be no less than 3.2m. A narrower driveway width (3m with access to both sides of the car or 2.6m with access to only one side) may be acceptable where the driveway does not have to provide a pathway to the house. Special consideration should be

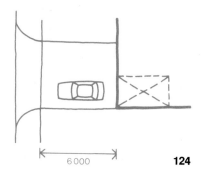

6 000   **124**

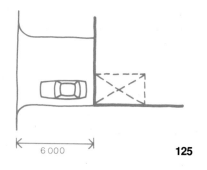

6 000   **125**

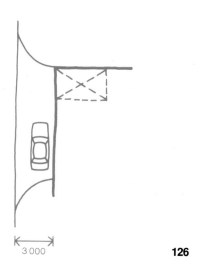

3 000   **126**

given to widths and lengths of driveways at dwellings designed for occupation by wheelchair users;

(c) have gradients below 10% or, whenever possible, no more than 12.5% for a distance 6m back from the edge of the carriageway.

Parking bays

3.90 As general guidance, it is suggested that grouped parking bays should normally have the following dimensions:

(a) when parallel with and contiguous with carriageways they should be 6m long by 2m wide or 2.4m wide where there is no footway or paved margin alongside (Figure 127);

(b) when at right angles to and contiguous with carriageways they should each be at least 4.8m long x 2.4m wide, and there should be 6m in front of the bays to allow for access and an additional 800mm strip at the back to allow for vehicle overhang (Figure 128). Such parking bays should normally not be provided along roads where more than around 100 dwellings are served.

(c) when in communal parking areas each parking bay should be at least 4.8m long x 2.4m wide. Bays for disabled people should each preferably be 3.6m wide or at least 3m wide where two adjacent bays may share an unloading area.

3.91 Space to accommodate trees should normally be allowed for in the layout of grouped parking bays and communal parking areas. Such trees will normally need to be protected to minimise risks of damage from vehicles, and species will need to be selected which do not create droppings that could cause damage to paint on cars parked beneath.

Communal parking areas

3.92 Dimensions for grouped parking bays in a 90 degree formation are given in Figure 129. The forecourt depth of 6m may be reduced to 5.5m by widening the parking bays from 2.4m to 3m. Parking bays with 60, 45 and 30 degree formations are less preferable and should no be used for culs-de-sac layout arrangements. Parking bay depths (D) and forecourt depths (A) for these formations should be as follows:

Formation	Dimension D	Dimension A
60 degree	5.4m	4.2m
45 degree	5.1m	3.6m
30 degree	4.5m	3.6m

3.93 Dimensions for grouped parking bays in parallel formations are given in Figure 130. The parking bay depth (D) may be reduced from 2.4m to 2m where the bay is bounded by a footway or a verge with a minimum width of 400mm (800mm if the verge is to be used for pedestrian access to cars). The forecourt depth (A) may be reduced to 3.5m for one-way traffic.

3.94 Forecourt dimensions between two rows of grouped garages are given on Figure 131. The forecourt depth (A) of 7.3m may be reduced to 6.5m when 3m wide garages (and correspondingly wider doors) are used. Additional length for turning at the end (C) should preferably be 3m or at least 1m.

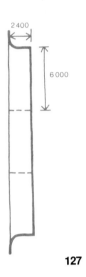

127

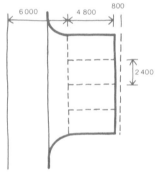

128

3.95 Dimensions required in front of a single row of garages are given in Figure 132. The forecourt depth of 6.8m is adequate when it is possible for vehicles to overhang a footway or a verge of at least 500mm when reversing.

3.96 As a general guide, it is suggested that driveways serving up to 25 grouped garages or hardstandings may be 3m wide, with 4.1m wide passing places where necessary. Driveways should be 4.1m wide when more than 25 spaces are served.

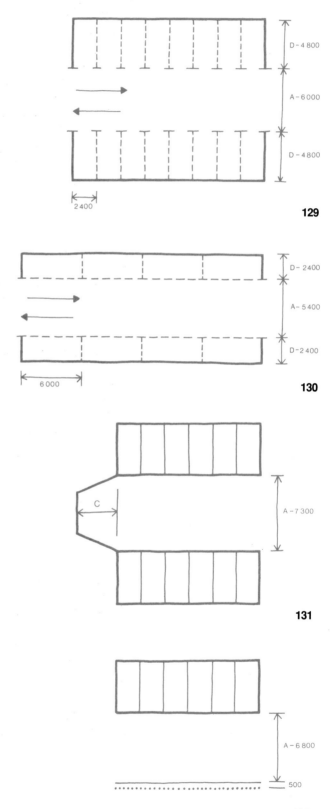

129

130

131

132

3.97 Vertical clearances should suit the vehicles to be accommodated in the parking area. Minimum clearances should be 2.1m in areas intended only for cars - with greater clearances at a change of slope. Other minimum clearances that may be relevant to the design of parking provision and access routes in high density developments are:

Small service vehicles	2.5m
Touring caravans	2.8m
Motor caravans	3.3m
Fire appliances	4.0m
Most large service vehicles	4.1m
The largest service vehicles	4.5m

3.98 Minimum headroom of 5.3m is recommended over public highways, for new construction, but maintained headroom should not be less than 5.03m, and for footbridges over public highways a headroom of 5.7m should be provided.

Demarcation

3.99 The configuration of the parking bay layout, changes in surfacing materials or rows of setts (or the like) should be used to demarcate individual hardstandings in grouped parking areas. Also:

(a) permanent and attractive means should be used to number grouped spaces assigned to individual households (not paint-markings on the surface or posts);

(b) parking spaces contiguous with carriageways should be demarcated with string courses in materials such as setts laid flush with the road surface;

(c) a change of surface material or a row of setts or paviors should be used to identify the boundary of adopted areas at entrances to private drives and driveways serving grouped parking areas.

4 Local Guidance and Standards

Residential road and footpath layouts

A corporate approach

4.01 Since Design Bulletin 32 was published almost all highway authorities have published new or amended guidance for the layout of residential roads and footpaths in new developments. Most authorities now take into account the principles that were set out in the bulletin. But there has been much discussion about how some of the detailed objectives and data should be interpreted in the preparation and application of local design guidance and standards. It is hoped that the new information presented in this edition will help to resolve some of these problems.

4.02 Experience since the first edition of this bulletin was published suggests that local design guidance needs to be produced jointly by the different professional disciplines within local authorities and in consultation with service providers, public transport operators, house-builders and other interested parties. Their joint aim should be to produce highway standards that enable designers to create visually attractive, safe, convenient, nuisance free and secure surroundings that are economical to construct and maintain.

4.03 Responsibilities for capital and maintenance costs are divided between housing developers, highways authorities and service providers, and a balance has to be struck between competing interests on these matters. For example, though an economical road layout can allow spaces to be used for building which might otherwise be wasted, the layout must also minimise damage and recurring maintenance charges that can be caused by vehicles over-running footways and verges.

4.04 To strike an appropriate balance involves taking a realistic view based on past experience of driver and pedestrian behaviour, and it is normally both unrealistic and uneconomic to plan for the worst imaginable behaviour and the worst possible combination of events. A balance must be struck, for instance, between the probability of damage being caused to footways by over-running vehicles, how obtrusive the damage would be, the savings in capital costs which could be made and the costs of repair compared with those of increasing the specification.

4.05 Local standards for road adoption which result from applying the advice in this edition will allow a sensible balance to be struck between planning, housing and highway considerations in the design of road and footpath layouts in the great majority of new developments. The use of more onerous standards will normally only need to be considered when there are special site circumstances.

Content of local guides

4.06 Development plans provide the framework within which the design of the residential road and footpath network should be set (see Section 1). Such plans will normally include special policies for conservation in areas of architectural and historic interest, rural areas generally and in non-designated urban areas where the existing character needs to be respected. Designs and layouts, and the application of standards for roads and footpaths, must be sufficiently flexible to preserve and enhance the character of these areas.

4.07 No guide can cover all the requirements that need to be taken into account when considering individual sites. Designers therefore need to know at the outset the context provided by development plans and other local policies and special constraints on access to the site or road and footpath layout requirements.

4.08 A corporate approach is needed to ensure that highway authorities standards are compatible with planning authorities objectives for residential density, layout form, the landscape and parking provision. Requirements should be included for all aspects of road and footpath layout that developers will normally be expected to take into account. Advisory standards should be distinguished from those which are mandatory.

4.09 The manner of presentation needs to be related to local circumstances. For example, standards may be set in a way that applies equally to a wide variety of design solutions to suit different kinds of urban, suburban and rural sites. Alternatively, standards may be related to the housing layouts most commonly submitted for approval, or to theoretical layouts in order to encourage innovation.

4.10 Prospective developers find it useful to know about schemes that authorities consider successful. Supplementary news sheets giving recent examples can be an effective way to do this. When proposed developments are large or in sensitive locations, it can be helpful if joint visits are made to relevant schemes by planners, highway and services engineers, emergency services advisers (police, fire and ambulance), bus operators and the developers.

4.11 Visits arranged for elected representatives can help to ensure that guides are understood and supported by members of highways, public transport, planning and housing committees.

4.12 Local authorities will need to apply their standards flexibly in relation to the principles and design data set out in this bulletin. And developers will need to use this information to produce well-reasoned design arguments if they seek the adoption of road layouts or details that differ from those envisaged in local guides.

4.13 Guides should encourage developers to obtain advice on matters related to:

(a) provision for services - from local offices of the public utilities and telecommunications industries;

(b) the provision of public transport services - from bus operators;

(c) crime prevention measures - from local police crime prevention or architectural liaison officers;

(d) the appropriate use of trees and shrubs - from sources such as landscape architects, horticulturists and arboriculturists.[59]

4.14 Guides should also describe the procedures and information required to obtain a road adoption agreement. For example, drawings will normally need to indicate which parts of the layout the developer expects to be adopted and how the adoption limits are to be demarcated on the ground. Widths and other key carriageway dimensions and the location and dimensions of parking spaces should also be shown.

Adoption agreements

4.15 Section 38 (and, less commonly, the Advance Payments Code under Part XI) of the Highways Act 1980 provide the statutory basis for the adoption of estate roads by mutual agreement.[60] House-builders and local highway authorities normally enter into such agreements so that, on completion to the authority's specification and satisfaction, the roads become highways maintainable at the public expense.*

4.16 There are obvious benefits from such agreements. House-builders are able to sell their houses more easily with road charges paid; authorities acquire roads constructed to their standards which should not require expensive maintenance and repair costs ahead of time; ready access to services is available in emergencies and for routine maintenance and house purchasers are assured that they will not be faced with road charges at a later time.

4.17 When completed to their specification and satisfaction, highway authorities should normally be prepared to adopt:

(a) access roads, shared surface roads, footpaths and cycletracks;

(b) land within visibility splays at junctions and on bends (to ensure that vegetation or other obstructions to visibility can be removed or trimmed if necessary);

(c) trees, shrubs and other features which are an integral part of vehicle speed restraints (see guidance in Section 1 regarding the adoption of trees and shrubs planted for other purposes);

(d) casual parking spaces provided for use by visitors immediately adjacent to carriageways (such spaces would not be allocated to residents and would be subject to highway law);

(e) service strips in unsurfaced ground immediately adjacent to shared surfaces, including strips that are located in front gardens and privately maintained.

4.18 Individual circumstances will need to be taken into account when considering the adoption of paved areas serving grouped hardstandings and garages, entrances and refuse chambers in blocks of flats and common facilities such as play areas.

Parking provision

Main considerations

4.19 Parking and other transport management policies in local plans will provide the basis for parking provision in local design guides. An important issue to be covered will be the provision of off-street parking for new development, and the plan policies will normally show the standards that the local authority is looking for.

4.20 Standards for the layout and location of parking spaces should be published - to ensure that developers are aware of requirements when preliminary layout plans are being produced and provide a common basis for discussions between developers and authorities officers.

---

* To help this process, a Model Section 38 Agreement has been produced jointly by the Association of County Councils and the House Builders Federation.[61]

4.21 Standards for the numbers of parking spaces to be provided should take into account the characteristics of typical new developments, distinguish between requirements for off-street and on-street provision, and be flexibly applied through the development control process to take into account:

(a) the location of the development (for this could affect the need for residents to own a car and the difficulties and costs of using it);

(b) the sizes and types of dwellings to be provided (for this could affect household size and composition and hence the number of car owners in the development);

(c) the proportions of grouped parking spaces assigned to individual households and the proportions of spaces provided within dwelling curtilages (for this could affect the extent to which the spaces provided would be available for general use).[62]

4.22 Section 2 describes the important role that layout design can play in helping to ensure that the parking spaces provided will in fact be used* (the location of off-street and on-street provision; the dimensions of spaces provided for vehicles to manoeuvre and park; the widths and depths of driveways, hardstandings and garages; and physical means to control the use of privately maintained and managed grouped parking spaces and shared driveways). Section 2 also refers to the role that planting and pavings can play in enhancing the visual character of parking areas. Section 3 gives dimensions for driveways, parking bays and communal parking areas.

Assignment of parking spaces

4.23 The total number of parking spaces required to meet residents demands will depend in part upon the extent to which the spaces to be provided will be assigned to individual dwellings. For example, at one extreme, the smallest total will normally be required when provision is to be made solely in the form of unassigned grouped hardstandings. In these circumstances, all of the spaces would be available for use by all residents and the total provided would only need to match the maximum number of cars that all of the residents together would be likely to own - taking into account differences in dwelling occupancy and differences in the numbers of cars owned by households occupying different sizes and types of dwellings in the development as a whole. The demands of households with above average car ownership would be set off against those with below average car ownership. Additional unassigned parking spaces would normally be required for casual callers and service vehicles.

4.24 In contrast, at the other extreme, the greatest total number of parking spaces for residents will normally be required when provision is to be made solely in the form of garages and hardstandings within dwelling curtilages. In these circumstances, the number of spaces provided within each curtilage would need to match the maximum number of cars that the different sizes and types of households that would be likely to occupy the dwelling would be likely to own. Consequently, requirements for the development as a whole would be the sum of these estimates of likely maximum parking demands for each size and type of dwelling. Additional unassigned spaces would normally be required for casual callers and service vehicles and to meet the needs of households who would find it impossible on some occasions to park within the curtilages of their homes.

---

* The consideration of other possible means lies outside the scope of this bulletin - for instance, waiting restrictions to help control on street parking.

4.25 When the layout has been planned to provide parking spaces on carriageways for casual callers, these spaces should be counted towards the total provision required for visitors cars and service vehicles.

Location of parking spaces

4.26 It is suggested that parking spaces intended to contribute to provision for residents cars should normally only be reckoned to include:

(a) garages, car ports and hardstandings within dwelling curtilages;

(b) grouped garages, car ports and hardstandings immediately outside the entrances of the houses or flats they are intended to serve - or, when there is no direct road access, at the ends of access footpaths.

4.27 It is also suggested that spaces for casual parking which are an integral part of the carriageway layout should normally only count towards provision for visitors cars and service vehicles if the spaces provided are immediately outside or in proximity to the dwellings they are intended to serve.

4.28 To arrive at an appropriate total requirement for a development, it will normally be necessary to consider each part of the layout separately - first to assess basic requirements for the accommodation proposed and then to make any adjustments that may be needed to suit the assignation of parking spaces and the proposed layout.

Improvement Schemes

The context

5.01 Incentives to improve the layout of existing residential roads and footpaths can be provided by area improvement policies such as the declaration of a renewal area or special projects such as group repair schemes, by policies to improve the quality of life on run-down public sector housing estates through renovation, and by policies of urban safety management such as the introduction of 20mph speed limit zones. Local authorities will need to take into account guidance on these policies in other publications.[63]

5.02 Though intended primarily for use in the design of new residential developments, the information and advice given in Sections 1-3 of this bulletin will provide a useful basis for assessing what improvements may be needed to existing residential roads and footpaths and for developing and appraising alternative design solutions. The special considerations that need to be taken into account when using these sections for that purpose are outlined below.

The improvement process

5.03 Relevant improvement policies and planning concepts in development plans will need to be identified at the outset. And the benefits and drawbacks of the existing roads and footpaths will need to be assessed with regard to matters such as their visual character, the quality of landscape maintenance, provision for access and parking, safety, convenience, security and patterns of vehicular and pedestrian movement within and immediately around the area.

5.04 Section 1 outlines the general considerations and evidence that will need to be taken into account when making such assessments. But the process will normally be more complex than that required to produce a design brief for a new development. Also, the assessments will be undertaken mainly by the local authority rather than by the house-builder. A corporate approach by the local authority will be essential at all stages in the improvement process to ensure that an appropriate balance can be struck between competing improvement objectives.

5.05 Residents and others such as the fire and ambulance services, bus operators, police (including the crime prevention officer) and local business people should always be given the opportunity to contribute to the assessments, and their views will need to be taken into account when identifying what improvements are needed. Individual households and representative organisations should be consulted at the outset and later when solutions are being developed. Proposed solutions should be presented pictorially to give a realistic impression of the intended alterations and additions. Residents views will be of special importance when determining priorities for action.

Assessments

Benefits

5.06 Residents and others should be given an opportunity to identify any existing benefits in their surroundings that they would wish to see preserved or enhanced. These benefits may include roads and footpaths which provide especially:

(a) convenient, safe and secure routes for pedestrians and cyclists;

(b) safe and secure places for children to play and residents to meet;

(c) quiet and nuisance free surroundings;

(d) visually attractive landscape views;

(e) convenient vehicular routes to destinations within and outside;

(f) convenient on-street and off-street parking provision.

5.07 Observations of pedestrian and vehicular movement, children's play patterns and the places where residents meet may be needed to complement such perceptions. Also, the visual character of the roads and footpaths and the condition of existing trees, shrubs and pavings will need to be described - paying special attention to features liked by residents. The overall design concept for the improvement scheme should aim to preserve and enhance the attributes identified.

5.08 An appropriate design concept will take into account residents opinions about the existing landscape, any relevant landscape proposals in development plans, and local policies for maintaining trees and plants on verges and incidental areas of common open space along roads and footpaths. Special efforts may be needed to identify places where new planting can be introduced.

5.09 It may be both appropriate and feasible to restore the original character of features such as front garden walls, hedges and gates, pavings and street furniture. Whenever practicable, the aim should be to design the road improvements so that any accumulated clutter of traffic signs and road markings may be removed.

Drawbacks

5.10 When asking residents and others to identify the drawbacks of the area it should be borne in mind that vehicular traffic may adversely affect satisfaction in the following main respects:

(a) accident risks (both actual and perceived) - caused by, for instance, excessive vehicle speeds, obstructions to visibility, inadequate carriageway alignment, inappropriate junction location or design, difficulties for pedestrians when crossing roads, poor street lighting and inadequate surfacing;

(b) noise - from tyres on the road surface, engines, exhausts, brakes and the bodies of heavy goods vehicles, many of which can be accentuated by braking and acceleration at junctions and by high speeds;

(c) vibration - from contacts between vehicles and the ground surface and from sound waves, causing, for instance, windows to rattle and floors to shake;

(d) pollution - from gases, lead and other suspended particles and black smoke caused by exhaust emissions;

(e) dust and dirt - especially from heavy goods vehicles;

(f) visual intrusion - from stationary and moving vehicles and street furniture such as traffic signs and signals and parking meters;

(g) severance - especially major roads separating residents from local community facilities;

(h) parked vehicles - especially parking on footways (a particular hazard to blind and partially sighted people) and the presence of heavy goods vehicles;

(i) damage - from vehicles over-running and parking on footways and verges.

5.11 Fuller descriptions of common problems can be found in the Transport and Road Research Laboratory's Urban Safety Projects reports[64] and in the IHT/DTp manual 'Roads and Traffic in Urban Areas'.[65] Problems in run-down public sector estates are described in the Department of the Environment's Handbook of Estate Improvement.[66] The Department of Transport's Manual of Environmental Appraisal[67] gives methods for measurement in order to determine the size and nature of problems.[68]

5.12 To complement residents perceptions of accident risks, local records will provide information on the actual locations and characteristics of any injury accidents which have occurred on roads within and adjacent to the improvement area. Site observations will normally be needed to record vehicle speeds. Noise levels, air pollution, indiscriminate parking patterns and damage caused by vehicles may also need to be recorded.

5.13 The views of residents and others about crime and vandalism in the area will need to be gathered, and any physical characteristics of the layout that cause concern - such as poor lighting and hiding places along the roads and footpaths - will need to be identified. When problems are caused by incidental areas of communal open space, residents views about alternative uses may also need to be sought (e.g. private gardens, play areas and allotments). Guidance on such matters is given in the Handbook of Estate Improvement.

Improvement potential

5.14 To help assess the improvement potential of the area it will normally be necessary to record vehicle, pedestrian and cyclist flows and the composition of vehicular traffic along a representative selection of roads in the area. And the road hierarchy within and around the area will need to be identified along the lines suggested in Section 1. The location of existing bus routes will also need to be recorded - to help identify any improvements which could be made to increase the attractiveness of bus travel.

5.15 The layout characteristics that will normally need to be identified and shown on drawings include:

(a) widths of carriageways, footways, footpaths and verges;

(b) kerb radii at junctions and bends;

(c) locations and dimensions of entrances to driveways;

(d) locations and dimensions of spaces that are available for parking on the carriageways or elsewhere;

(e) distribution of statutory and other services underground;

(f) positions of surface water drainage gulleys;

(g) positions of lighting columns and other street furniture;

(h) locations and conditions of trees and other plants;

(i) locations of bus shelters and stops.

5.16 It will also normally be useful to indicate on drawings the locations of accidents and accident risks, the places where vehicles are actually parked during the daytime and at night and the main places where children play and residents meet.

5.17 To help assess the likely consequences of altering the road layout it may be sensible in the first instance to experiment by introducing changes that are economical to install and that could later be either readily reversed or, with additional expenditure, made permanent.

Costs and benefits

5.18 Assessments of the costs and benefits of alternative design solutions will normally be needed to help set priorities for expenditure between individual improvement items (and between one improvement area and another). Costs include those of implementation, enforcement and maintenance - and those for road users. Benefits include those arising from reduced road maintenance costs, fewer accidents, savings from reduced congestion, reduced opportunities for crime and vandalism and improvements in environmental quality.

Design objectives and solutions

Scope

5.19 The layout design objectives for new housing in Section 2 will mostly be relevant when considering improvements to existing roads and footpaths. For instance, it will usually be important to find ways of discouraging non-access traffic and restraining vehicle speeds whilst providing convenient vehicular routes for residents and those who provide services.

5.20 Many of the design solutions in Section 2 will also be relevant. But existing arrangements of buildings, roads and footpaths - and a scarcity of resources - may limit the extent to which the full range of options can be considered. And the scale and nature of traffic problems in some improvement schemes may require special solutions which would be inappropriate for most new developments.

Non-access traffic and vehicle flows

5.21 An area-wide approach is needed when determining what action to take to minimise danger and nuisance from non-access traffic and excessive vehicle flows. Patterns of movement, accidents and environmental problems over a wide area have to be considered when determining the boundaries of an improvement area. And it may be necessary to consider whether roads in the wider area ought be improved. For example, selective road closures may be needed to reduce the numbers of junctions along distributor roads in order to improve safety and traffic flows along the distributors and thereby help to exclude non-access traffic from the improvement area.

5.22 Conversely, improvements to roads within the improvement area may help the local authority to work towards a hierarchical structure for roads over a wider area. Speed restraints along residential roads will often cause sufficient inconvenience for drivers to ensure that non-access traffic keeps to the distributor roads - thus making it unnecessary to close residential roads or change priorities at junctions within the improvement area. The use of one-way streets will not normally be appropriate in improvement areas that are primarily residential.

5.23 Desirable maximum vehicle flows in different parts of the improvement area will often vary according to the physical characteristics and functions of the roads. Points of access to the area and alterations to the road layout should be designed to avoid creating excessive flows.

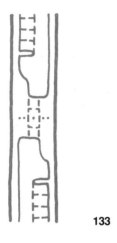

133

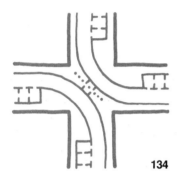

134

5.24 It will not normally be either sensible or feasible in improvement areas to remove existing accesses to dwellings from roads serving in excess of around 300 dwellings (the threshold for avoiding direct access to dwellings that was suggested for new developments in Section 2).

5.25 Turning bays should be created at the ends of culs-de-sac (or facilities should be available at nearby junctions) to ensure that vehicles do not have to reverse over long distances. Also, it should be possible to link turning bays with each other (Figure 133) or with other roads to provide alternative means of vehicular access in an emergency and to retain through routes for pedestrians and cyclists.

5.26 Unless significant safety problems could arise, cycle gaps should always be provided at road closures such as that illustrated in Figure 133 and also at junctions which have been altered to cut off the through routes - such as the cross road illustrated on Figure 134.[69] In some circumstances it may be necessary to provide barriers that will allow access by cyclists but prevent motor cyclists passing through. However, such barriers should only be used if there is a significant problem - because they inconvenience pedal cyclists by requiring them to dismount.

Vehicle speeds

5.27 The guidelines for speed restraints set out in Section 2 will need to be applied with special care and flexibility in existing residential areas - taking into account not only matters such as the physical characteristics of the roads and patterns of on-street parking but also the location of any accidents which have occurred in the area. Full consultation is essential with local residents and others who have an interest such as bus operators and emergency services.

5.28 Subject to the considerations outlined above, road closures and junction alterations (including changes in priority) may be used to break up long lengths of roads and thereby help to restrain vehicle speeds. Islands, refuges and some types of narrowings and chicanes (Figure 135) may serve the same purpose when existing carriageways are wide enough and indiscriminate on-street parking would not create blockages. However, it will often be necessary to complement such restraints by using road humps - especially raised junctions and speed tables. In some areas it may be necessary to use changes in vertical alignment alone.

135

5.29 As suggested in Section 2, raised junctions may be used to maintain low speeds and provide improved crossing facilities (Figures 136 and 137). Also, series of speed tables may be used on long stretches of road and to complement carriageway narrowings created by parking bays and footway extensions. The latter extensions will provide the only opportunities that are available to plant trees in many existing streets.

136

137

5.30 The Highways (Road Hump) Regulations 1990 permit humps to be used along roads with a 30mph speed limit, provided the proposed arrangement and signing conform to the requirements set out in the Regulations.[70] It should be noted that the Regulations allow flat top and round top humps, both of which may be kerb to kerb or tapered - the latter being advantageous because existing drainage arrangements are not affected. A kerb to kerb hump would need to be provided at a pedestrian crossing place.

5.31 Humps may be used in 20mph zones without many of the restrictions which apply to their use on other roads (though these humps - like any other speed restraint - must be clearly visible to drivers and others during the day and at night). Highway authorities will need to be aware of the design and other requirements that have to be met before a 20mph speed limit can be introduced.

5.32 The IHT guidelines on Urban Safety Management discuss ways of ensuring that non-access traffic uses appropriate roads in the hierarchy.[71] The remaining roads may be suitable for the introduction of speed reducing measures in association with a 20mph speed limit.

Provision for pedestrians

5.33 Improved layout arrangements will need to be considered at places where pedestrian safety is at risk and where there are barriers which restrict the movement of wheelchair users and other people with disabilities. For instance, new pedestrian crossings or islands may be needed on adjacent distributor roads, and the measures required to exclude non-access traffic and restrain speeds may need to be specially directed at improving facilities for pedestrians in specific places (such as at junctions). The removal of obstructions that restrict footway widths and the introduction of dropped kerbs are amongst the improvements that may need to be considered to assist people with disabilities.

5.34 The costs involved in turning existing streets into shared surfaces are normally substantial, and the resources involved will usually be better spent on making other improvements to create safer conditions for pedestrians. However, it will normally be entirely appropriate to create a shared surface where the carriageway is too narrow to accommodate both moving vehicles and parked cars, and the footways are used for parking. It will normally only be possible to overcome these problems by utilising the whole area occupied by the carriageways and footways (Figure 138).

138

Provision for cyclists

5.35 Where suitable quiet roads are not available for cyclists as an alternative to the busier and less safe roads, measures may be required to improve safety at particular locations. Additionally, the conversion of footways to segregated shared use by both pedestrians and cyclists may need to be considered. Policy advice and technical information on shared use is contained in Local Transport Note 1986/2.[72]

5.36 Cyclists are best segregated from pedestrians by means of a physical separation, such as a raised kerb (with pedestrians stepping down to the cycle track). Where this is not possible, segregation by a continuous white line has been shown to be effective. For the latter situation it may be necessary to consider the needs of visually impaired people and provide tactile slabs at suitable locations so that they can be made aware of the area.

5.37 On occasions it may not be possible to provide separate facilities for both pedestrians and cyclists, e.g. because of the limited footway width available. In these cases it is possible to allow pedestrians and cyclists to share the same area by converting the whole width of the footway to a cycle track. When possible, a verge between the cycle track and carriageway, or some other form of protection, should be provided where minimum standards are utilised.

5.38 Cycling on footways or footpaths is illegal. Powers contained in the Highways Act 1980 allow all or part of a footway to be converted to cycle use as a cycle track: Section 3 of the Cycle Tracks Act 1984 provides a procedure for converting all or part of a footpath to a cycle track. Section 2 of the 1984 Act makes it an offence to drive or park a motor vehicle (including a moped) on a cycle track, except where specifically exempted.

5.39 Detailed information on the conversion of footways or footpaths to cycle tracks, the creation of new cycle tracks, appropriate signing, cycle crossing facilities and the planning and development of cycle routes can be found in guidance published by the Department of Transport.[73]

Parking

5.40 Parking demands in excess of the provision available commonly causes indiscriminate parking in improvement areas, and such parking may prevent vehicular access to dwellings, create problems for the emergency services and bus operators and damage footways and verges. Existing patterns of roads and buildings normally limit what can be done to increase or improve provision for off-street and on-street parking - thus simple solutions to the problems created by parking are often not available. Both traffic management and design measures will normally be needed. Any local policies regarding traffic restraint or policies regarding conversions and changes of use for buildings in the area will also need to be taken into account.

5.41 The general duties of the traffic authority are contained in Section 122 of the Road Traffic Regulation Act 1984.[74] These are relevant to the provision of on-street and off-street parking facilities, and may lead to the use of parking orders to establish on-street residents' parking spaces. The likely effects on the surrounding district of any displacement of parking which would result from such an improvement would need to be taken into account.

139

5.42 In some schemes, parking provision may be increased by using land occupied by unnecessarily wide footways, verges and carriageways (Figures 139 and 140), large front and rear private gardens and little used areas of common open space. Speed restraints and other road improvements should be designed and located to minimise reductions in numbers of parking spaces.

5.43 Indiscriminate on-street parking can cause special problems for pedestrians. Such parking can make it difficult for pedestrians (particularly children) and drivers to see each other. Thus, it may be necessary to widen the footway in places to allow pedestrians to see beyond parked cars before stepping out onto the carriageway (as illustrated in Section 2). Trees may be planted in these places and also where footways are extended in a more modest way to indicate and control the location of on-street parking spaces.

5.44 Footways may also need to be extended to prevent vehicles parking on the carriageways and obstructing visibility at junctions. Permanent barriers such as high kerbs, bollards or railings may need to be provided to prevent parking on footways in these places and also at junctions with footpaths and where there incidental planted areas adjacent to the carriageway.

5.45 The design data in Section 3 will be of use when planning such improvements in detail. But the dimensions suggested are intended for new developments and would need to be applied flexibly in the context of improvement schemes.

140

Appendix 1

Geometric characteristics of vehicles turning

1. Figures 1-5 show the wheel tracks and body overhang for each of the vehicles listed in Appendix 1 of the first edition of this bulletin turning through 90 degrees. They are based on the formula in Appendix 1 of the Transport and Road Research Laboratory report 608 [75] to which designers may refer for additional information. Figures 2-5 show the vehicles turning at low speed in full lock. Figure 1 shows the widths of the wheel paths plus body overhang for different radii of turn.

2. Figures 6-29 show the paths described by the same vehicles making various turning manoeuvres. The paths are based on studies conducted by the TRRL* for the first edition of this bulletin, and represent the typical movements made by vehicles turning at speeds of between 5-10mph.

3. All figures are to a scale of 1:200. These may be used by designers when considering the form and dimensions of junctions, bends and turning spaces. It must be emphasised that all the dimensions in Figures 1-29 are the minimum required for vehicles to complete their manoeuvres. While some tolerance is implicit in so far as each vehicle is the largest of its type, clearances between vehicles and kerbs will be required when the data are applied.

1 Outside turning radius of front axle (m)

15		30		45		60		75-400		400 +		
X	Y	X	Y	X	Y	X	Y	X	Y	X	Y	
3.44	3.89	2.96	3.19	2.80	2.95	2.73	2.84	2.68	2.77	2.53	2.54	Pantechnicon
2.94	3.33	2.66	2.85	2.58	2.71	2.53	2.63	2.50	2.58	2.42	2.43	Refuse vehicle
2.67	3.06	2.42	2.61	2.34	2.47	2.30	2.40	2.27	2.35	2.19	2.20	Fire appliance
1.96	2.10	1.84	1.91	1.80	1.85	1.78	1.81	1.76	1.78	1.73	1.74	Private car

X = Maximum width of wheel path

Y = Maximum width of wheel path plus overhang

---

*Conducted by the Traffic Systems Division of the Traffic Engineering Division TRRL.

Vehicles turning through 90°

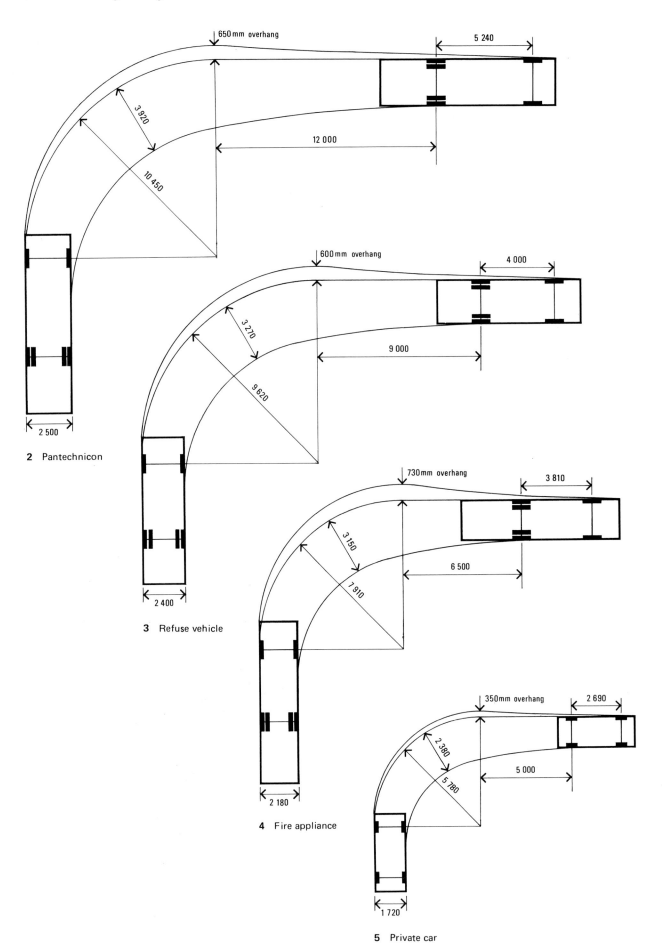

650 mm overhang

5 240

3 920

10 450

12 000

2 500

2 Pantechnicon

600 mm overhang

4 000

3 270

9 620

9 000

2 400

3 Refuse vehicle

730 mm overhang

3 810

3 150

7 910

6 500

2 180

4 Fire appliance

350 mm overhang

2 690

2 380

5 780

5 000

1 720

5 Private car

Full lock forward

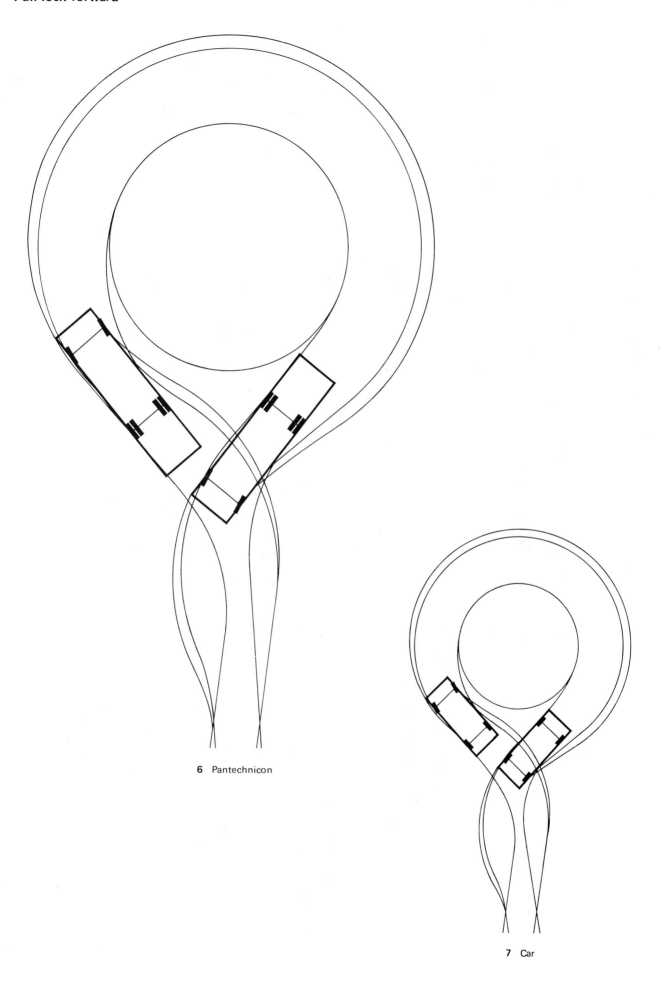

6 Pantechnicon

7 Car

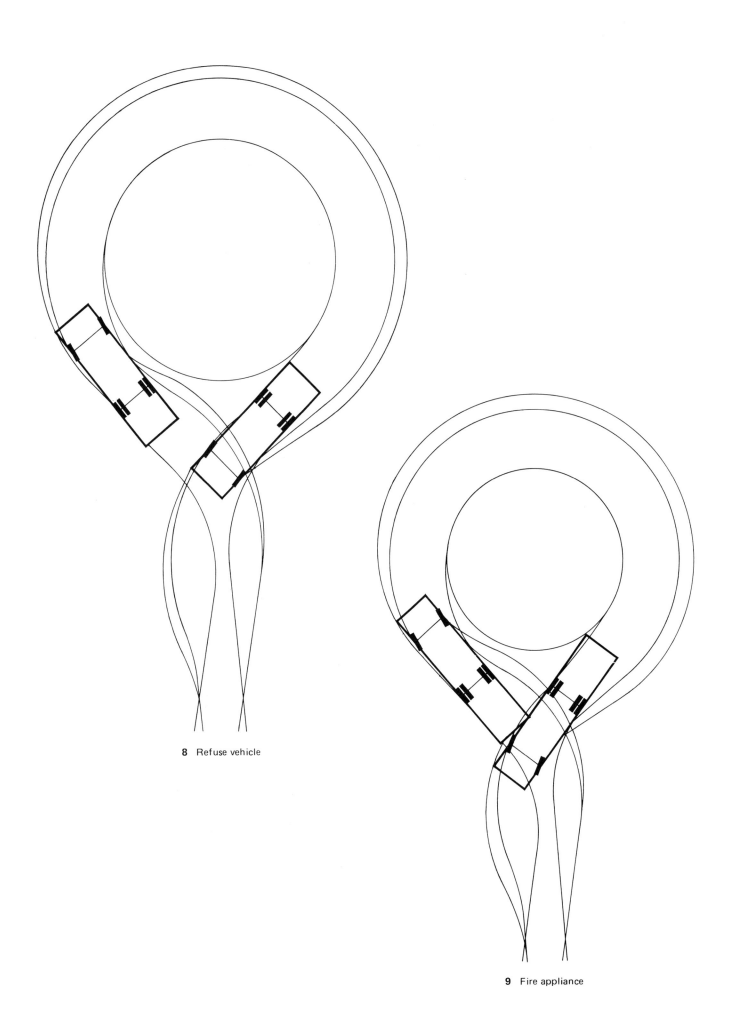

8 Refuse vehicle

9 Fire appliance

Full lock reverse

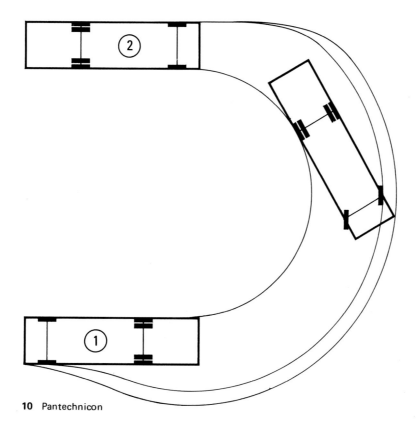

10 Pantechnicon

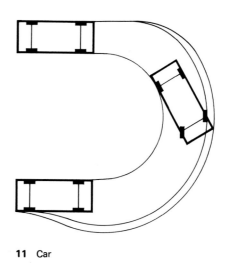

11 Car

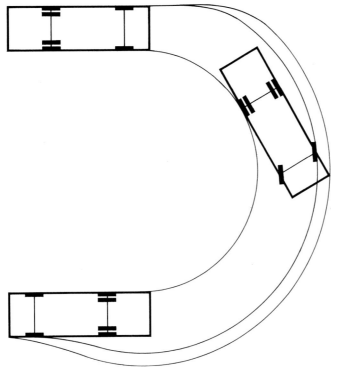

12 Refuse vehicle

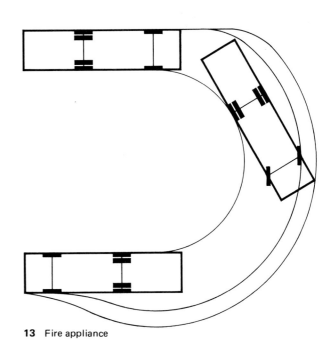

13 Fire appliance

Hammerhead — T form

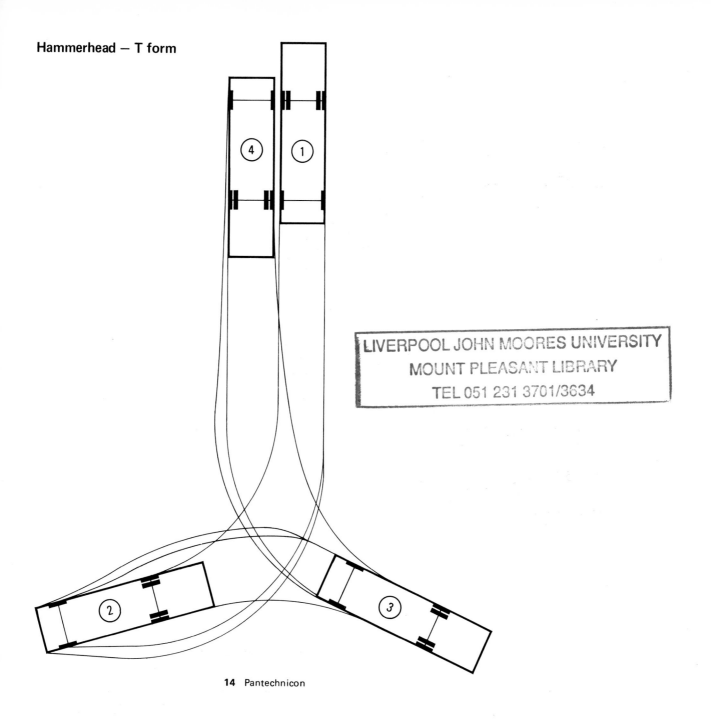

14 Pantechnicon

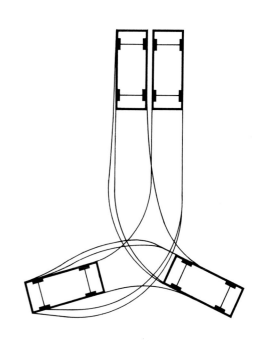

15 Car

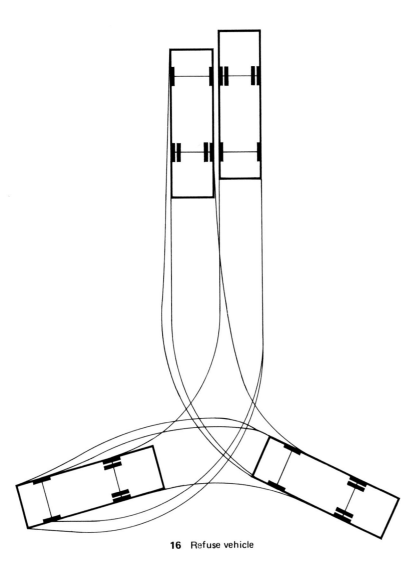

16 Refuse vehicle

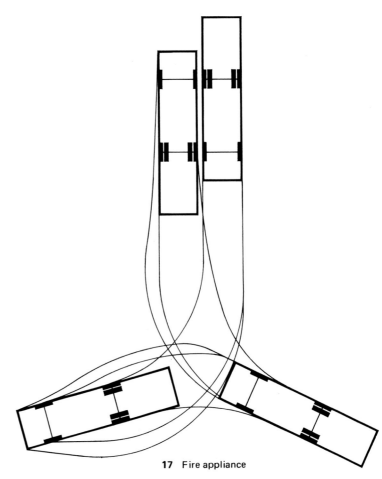

17 Fire appliance

Hammerhead — Y form

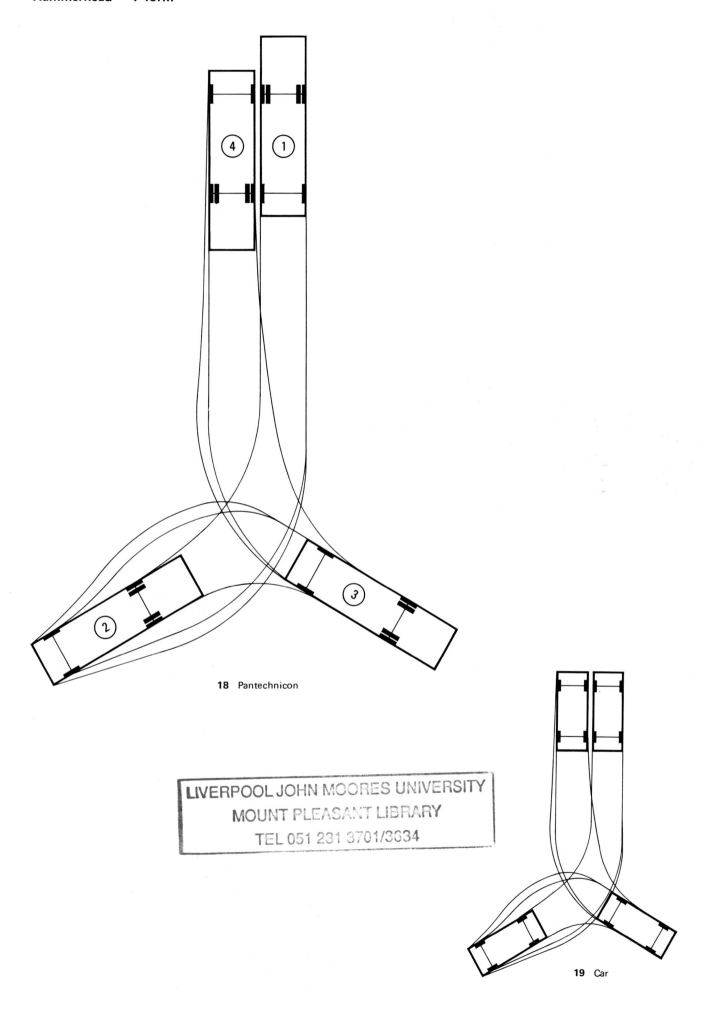

18 Pantechnicon

19 Car

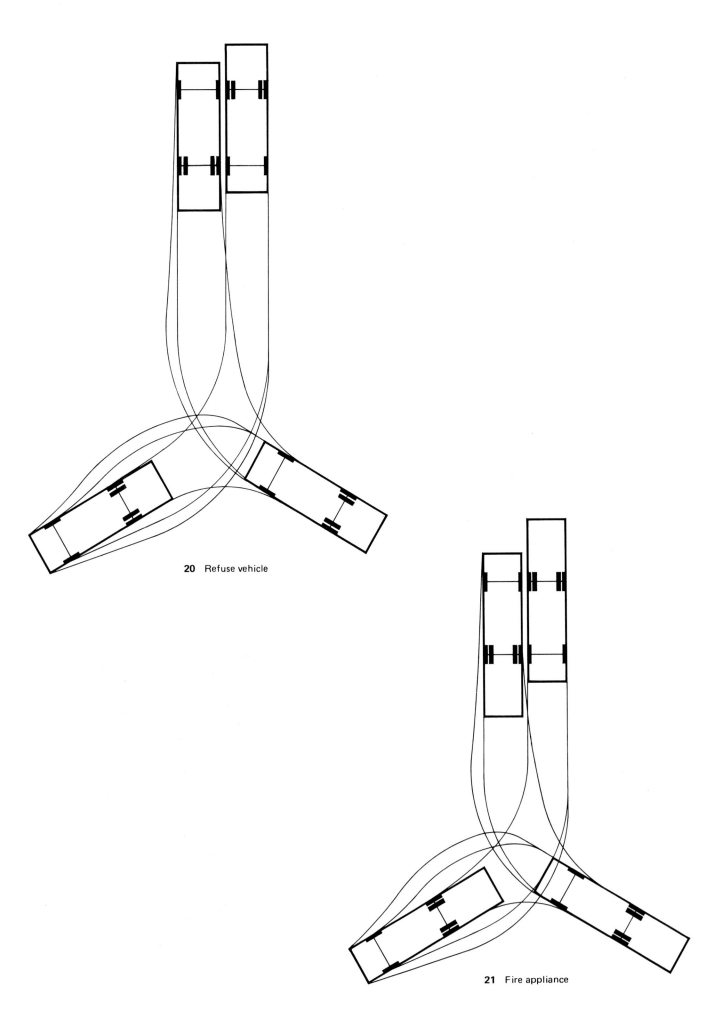

20 Refuse vehicle

21 Fire appliance

Forward side turn

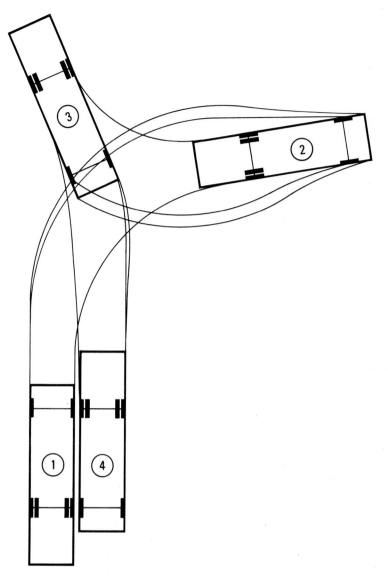

22 Pantechnicon

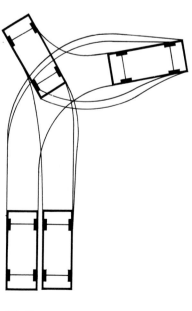

23 Car

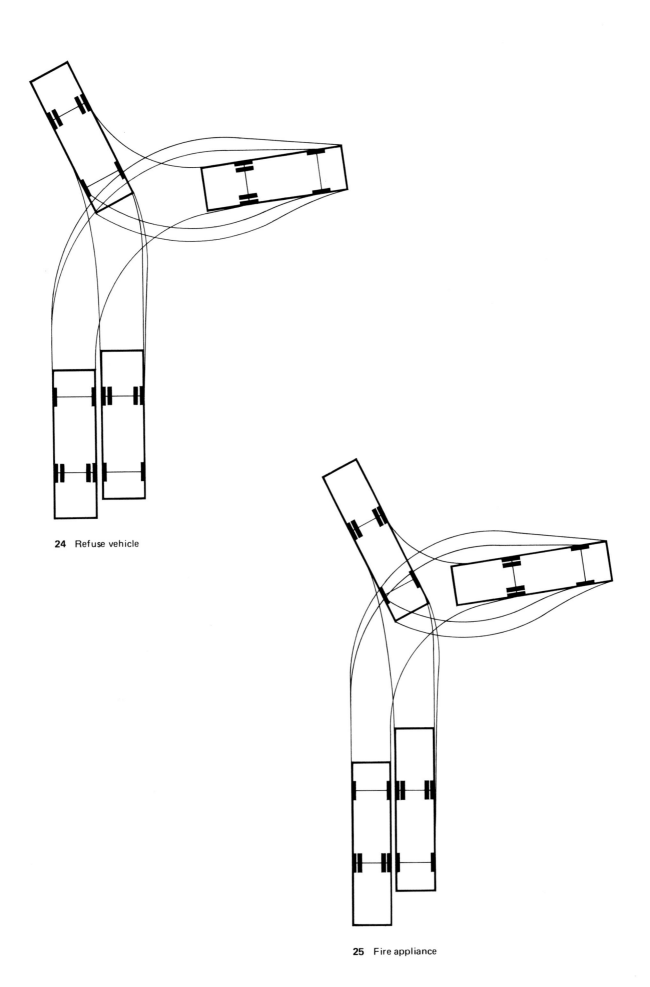

24 Refuse vehicle

25 Fire appliance

Reverse side turn

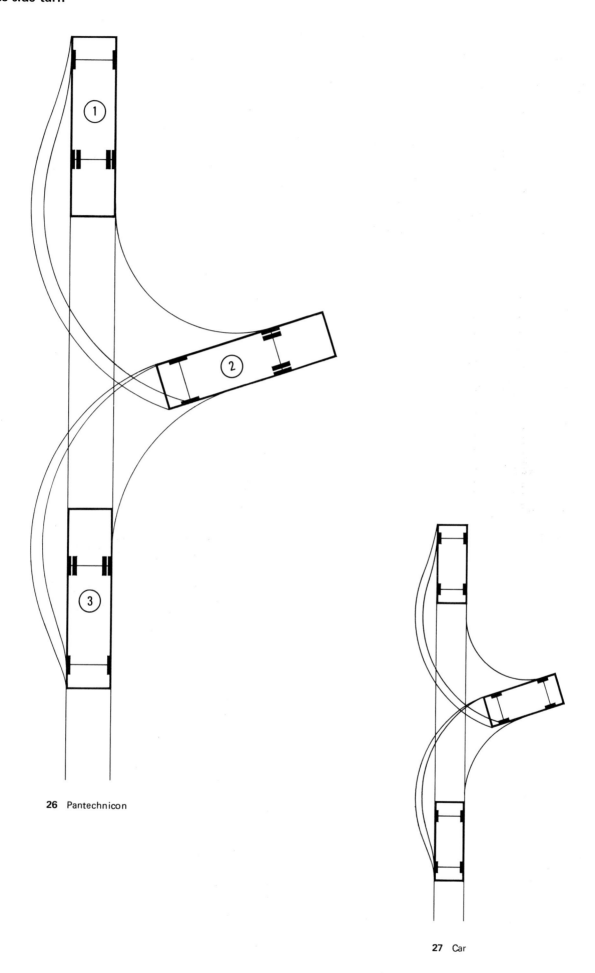

26 Pantechnicon

27 Car

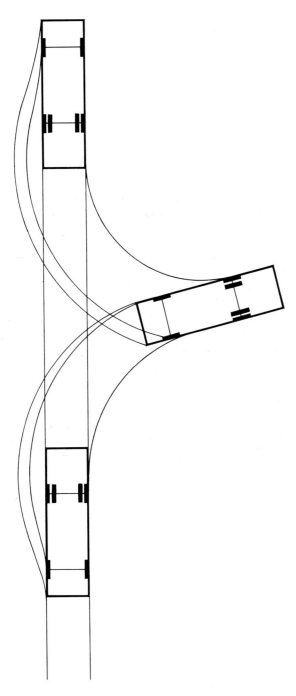

28 Refuse vehicle

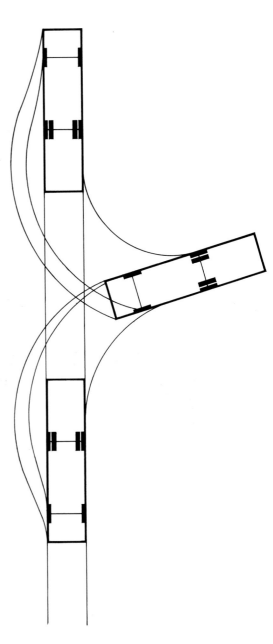

29 Fire appliance

Passing places on narrowed carriageways

1. To avoid opposing vehicles confronting each other on a narrowed stretch of carriageway they should be able to see each other before a passing bay is reached by the vehicle nearest to it.

2. Thus in Figure 1 for vehicle B to pull into the bay it is necessary for the driver to see vehicle A before B has reached point O. Point O is dependent on the speed of vehicle B and is the point beyond which B cannot slow down or stop in time to enter the passing bay. But, because either vehicle A or B may be closest to the passing bay, both vehicles must be able to see each other before either has passed the point where it cannot slow down or stop in time to use the bay.

3. The implications of this are shown in Figure 2, illustrating that with a sequence of passing bays the forward visibility distances overlap, each being determined by the combined stopping distances of opposing vehicles plus the distance required between bays necessary to cope with the traffic volumes envisaged (see Appendix 4 in the first edition of this bulletin Studies of delays to traffic on single-lane carriageways with passing places[76]).

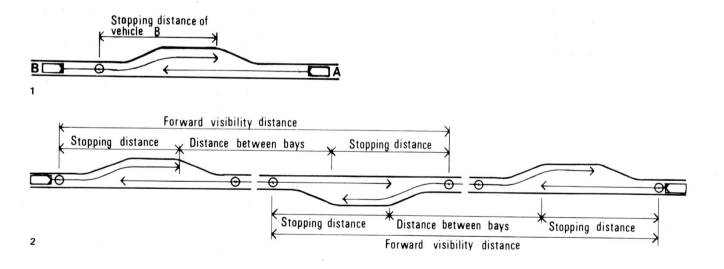

4. Where bends occur, the same rules will apply as shown in Figure 3 and it will be appreciated that if passing bays were not provided on the bend the sight lines required between bays on each side of the bend could sterilise a comparatively large amount of land. For similar reasons, it will also normally be necessary to provide passing places at junctions between narrowed carriageways (Figure 4) and at their entry point from carriageways of normal width (Figure 5).

5. It should be emphasised that Figures 1-5 are purely diagrammatic. The shape and size of passing bays required will depend largely on the types and volumes of traffic to be coped with. Their location and design at junction points may also be affected by the need to allow vehicles to turn past others waiting in the passing bays. This again will be influenced by the volume of traffic but may also be influenced by the direction of turn at the junction point (see Section 3).

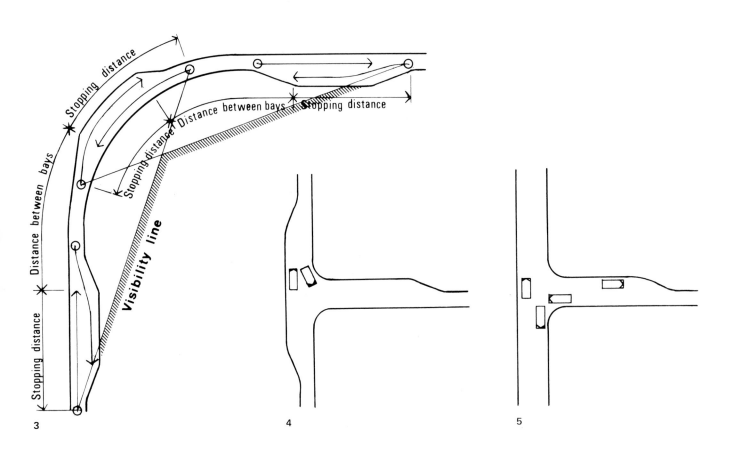

3

4

5